11/18

TEN

R wanee

Books should be returned or renewed by the last
date above. Renew by phone **03000 41 31 31** or
online *www.kent.gov.uk/libs*

599·658

REINDEER

REINDEER

An Arctic Life

Tilly Smith

To our grandson Hamish.
The future is with the young.

First published as *The Real Rudolph: A Natural History of the Reindeer*, 2006.
This fully revised and updated edition first published 2018

The History Press
The Mill, Brimscombe Port
Stroud, Gloucestershire, GL5 2QG
www.thehistorypress.co.uk

British Library Cataloguing in Publication Data.
A catalogue record for this book is available from the British Library.

ISBN 978 0 7509 8797 4

Typesetting and origination by The History Press
Printed in Turkey by Imak

CONTENTS

PREFACE

I AM A SELF-CONFESSED reindeer geek. I have a house full of reindeer photos, paintings, antlers, skins, books, chopping boards and other reindeer memorabilia I have collected. Many of my foreign holidays have been 'busman's holidays' to reindeer herding regions and barely a day goes by when I am not doing something related to reindeer. I am also lucky enough to be co-owner of Britain's only free-living herd of reindeer – the Cairngorm Reindeer Herd.

My interest in deer and subsequent 'love' of reindeer brought me to work as a volunteer with the Cairngorm reindeer in the summer of 1981. I never looked back – thirty-seven years on I am still in my first job (I eventually got paid!) and have enjoyed every minute of it. I regard myself as an extremely lucky person whose passion is their daily work.

So when I was asked to write this book I was delighted to accept and confident I had the credentials. But writing books takes time and so I am immensely grateful to my family and work colleagues who have kept the ark afloat while I pontificate about reindeer.

To avoid confusion it is important to point out right away that reindeer and caribou are interchangeable names for the animal species *Rangifer tarandus*. Throughout the book both names are used depending on the region of the world I am referring to. In Alaska and Canada, caribou is the commonly used term and in the Scandinavian countries and Russia the same animals are called reindeer.

· 1 ·

WINTER

THE WEATHER FORECAST for the Cairngorm mountain area, Britain's only sub-Arctic habitat in the UK, is 'winds gusting 100 miles per hour, serious wind chill factor and arduous walking conditions'. These are severe winter conditions with drifting snow, and the Cairngorm Mountain Rescue team have been called out to look for two lost hill walkers; their prospects are not good.

While a full-blown search is being mustered to find the missing walkers, the Cairngorm reindeer are out there too, probably lying down, chewing the cud and feeling quite at home as they wait for the wind to abate. Their antlers will be frosted with blown snow, any snow lying on their backs will easily shake off when they stand up and their bed of snow is a comfortable one. In Alaska, one of the countries where caribou are naturally found, they say there are only two seasons, 'snow' and 'no snow', and caribou thrive there. They are truly Arctic animals, totally at home in the coldest places in the world.

From the tips of their noses to the bottom of their feet, reindeer are covered in hair. In addition, the quality and density of a reindeer's coat are unrivalled in their suitability for this climate. The coat is made of two types of hair. Hollow guard hairs, composed of a matrix of air-filled cells, are anything up to 4–5cm long and are densely packed.

The hair root itself is thin but outside the skin the guard hair expands immensely. Below this is a fine woolly coat, not air-filled but much denser. The combination and density of the two types of hair provides the reindeer with fantastic insulation. In fact, their coat can be likened to an extremely high-tog-value duvet filled with hollow fill or feather down, air of course being one of the best forms of insulation.

The two-layered coat of reindeer is incredibly dense: 670 hairs per cm^2 for the longer hollow hair and 2,000 hairs per cm^2 for the woolly undercoat. In the past reindeer hair was used extensively for stuffing upholstery and even spun with wool to make clothing. Unlike sheep's wool, the staple of reindeer hair is not long enough to be spun on its own.

They are so well insulated that reindeer can lie on snow without melting it, so a layer of snow on the ground provides a comfortable dry bed. Also, snow that lands on their backs doesn't melt – it remains frozen and can itself add to the insulation value. A reindeer is like an enormous thermos flask wandering around, its coat letting no heat out and no cold in.

Boots made out of reindeer skin were the only footwear that kept my feet warm in −35°C Swedish Lapland. And they weren't any old reindeer boots

either but Russian ones, which have two layers of reindeer skin, with hair facing in the way and out. With no hard sole, I felt as though I was walking about on the snow with my slippers on, but in this dry, cold environment reindeer skin boots are the business.

It was the Jokkmokk Winter Market, a traditional festival that takes place in February every year and we were at the 300th one. Jokkmokk is situated right on the Arctic Circle and this small town is in the middle of Swedish Lapland. The market is a great celebration of the Sami culture and their reindeer, and is a must for any reindeer enthusiast, as long as you are happy to embrace reindeer alive and dead. The sleepy town is transformed into rows and rows of stalls selling anything from reindeer meat to outdoor clothing. There is reindeer racing and displays of Duodji, a meticulous and incredibly expensive Sami handicraft in which reindeer antler has the most intricate patterns and designs skilfully scored onto the surface. The decorated antlers make beautiful knife handles and they are exquisite. But the highlight of each day is the reindeer parade. A train of white reindeer, each one pulling a traditional Pulka (a boat-shaped sledge) and led by a Sami reindeer herder in full dress, from his ornate hat to his reindeer skin boots with turned-up toes. As the procession walks through the town, the crowds gather to get the best view; it is a wonderful display of Sami tradition and their docile Arctic animals.

Away from the market in the surrounding forests there are reindeer everywhere, searching out an existence by digging down through the deep snow in search of food. Often all you can see are reindeer bottoms as they are head first in the deep snow, their large hooves shovelling down to the vegetation below. Winter offers an extremely limited menu; reindeer have to resort to an impoverished diet, which for the vast majority means lichens. Lichens are very interesting as they actually consist of two different organisms in a symbiotic relationship. The main body is made up of a fungus, which grows like other fungi by living on a supply of organic food. Within the fungus live unicellular algae, which contain chlorophyll and grow by the process of photosynthesis. The fungal component and the algae all live on the food manufactured by the algae. Poor in protein but rich in carbohydrate, lichens provide reindeer and caribou with instant

In some areas lichen will comprise as much as 80 per cent of the reindeer's total winter diet. The lichens generally preferred by reindeer are the mats of ground lichens, particularly fruticose lichens, including *Alectoria, Cetraria, Cladonia* and *Stereocaulon*.

energy but an overall negative nitrogen balance. In other words, reindeer survive but don't grow while on this lichen diet, steadily losing condition over the winter.

Their hair extends right down to their lips – an extremely important requirement in Arctic conditions and no doubt hardened polar explorers grow beards for the same reason. Breathing out warm air into sub-zero conditions creates a build-up of frost on any cold wet surface, like a wet nose. By having a completely hairy muzzle reindeer avoid this and instead can go about their everyday business with warm, dry noses.

> Reindeer and caribou are not alone in eating lichens. Black-tailed deer and Chinese musk deer also feed on them, along with spruce grouse and wild turkeys.

Extremities such as the legs are more vulnerable to heat loss. However, a specialised arrangement of blood vessels going to and from the legs helps to reduce this problem. Warm blood flowing to the legs passes closely by the cold blood returning from the legs. This counter-current system allows heat and cold exchange between the two so that the warm blood going to the legs is cooled and cool blood leaving the legs is warmed, therefore reducing heat loss from the legs.

A similar heat-exchange system operates in the nasal passages of the reindeer. They have particularly complicated nostrils with bone and cartilage designed like a 'rolled scroll'. This, combined with many fine, short hairs, creates a highly effective method of retaining heat and water. By greatly increasing the surface area of the nostrils, blood can warm the cold incoming air. This enables the water in the air to condense and it then trickles back into special folds, which direct it to the back of the nose and into the throat. This nasal heat exchange protects the reindeer from heat and water loss when they breathe in the cold. As the warmed air then travels down the windpipe to the lungs the neck is kept warm by the long hollow hairs or 'beard' below the neck.

Swedish scientists have used thermal imaging cameras to show how the main body of the reindeer is almost 'invisible' with minimal heat loss from the legs. However, they were surprised how their noses glowed! This is due to a high concentration of blood vessels keeping the nose and lips warm and sensitive, presumably so they can detect what they are actually eating through the thick layers of snow. I suspect a nose numb with cold would be as ineffective as numb fingers!

There is an old Russian saying: 'There is no such thing as bad weather, merely unsuitable clothing' – a great phrase that sums up reindeer in winter.

• 2 •

REINDEER
AND CARIBOU

REINDEER AND CARIBOU (*Rangifer tarandus*) are the only Arctic-living species in the deer family (Cervidae), a group that includes at least forty different species of deer spread across the world and occupying a vast range of ecological niches. The smallest member of the family is the Pudu (*Pudu mephistophiles* and *Pudu puda*), which frequents the forested slopes of the Andes, while the largest is the moose (*Alces alces*), which inhabits the forests and marshlands of northern Europe, Siberia, Canada, Alaska and a few states in North America.

Found in the tundra, mountains and woodlands of the Arctic and sub-Arctic areas of the northern hemisphere, reindeer are in many ways unique. Many herds migrate further than any other deer species: as much as 1,000km between their summer and winter range is not uncommon. They are often highly gregarious, forming huge herds of many thousands, and survive primarily on lichens in the winter. Their antlers are asymmetric and are grown by both sexes, including calves from just a few weeks old. They are also the only deer species where there is a wild and domesticated form.

Reindeer are quite heavily built and stocky, compared to other deer. Red deer portrayed by Victorian artist Landseer as 'The Monarch of the Glen' are slim, elegant and have a rather proud posture, holding their heads high, with long legs and pointed hooves. Reindeer carry their heads very low to the ground; with a chunky profile, thick legs, short hairy ears and flat feet, they have an ungainly look about them. But by having a low body-to-surface area ratio reindeer lose relatively less heat than their cousins.

Compared with all of the other species of deer, reindeer and caribou have relatively large feet for the size of their bodies. Their hooves are concave with sharp edges and their back dew claws are also large. Both the dew claws and the main cloven hooves can be spread far apart. Splayed hooves spread the weight of the reindeer across a larger surface area, and so the weight load on the ground from individual feet is low. In their travels reindeer regularly encounter soft, deep snow and wet, boggy ground at different times of year, so spreading their weight to allow them to travel with little effort is vitally important.

The mature bulls grow big 'showy' antlers, the largest among the deer species relative to body size, whereas the cows grow smaller more practical antlers, which give them considerable status in the herd during winter. Supremely adapted to their hostile environment, reindeer are a classic Ice Age mammal and one of the few large mammals to survive the rise in temperature after the last Ice Age, 11,000 years ago. Today, reindeer range over one-fifth of the earth's surface and the total world population is approximately 6 million.

In 2015, wild reindeer were categorised as Vulnerable A2a by the IUCN Redlist due to a 40 per cent decline in their overall numbers across their whole range, which is approximately 3 million reindeer.

Tuktu is the Inuit name for caribou, but another school of thought believes that reindeer are named after the Lappish word *'reino'* meaning calf.

This figure includes all wild and domesticated reindeer. The name reindeer is derived from the Old Norse word *hrein*, meaning reindeer, while in North America caribou is derived from the Micmac Indian word *Xalibu*, meaning 'digger of snow'.

Without getting caught up in lots of taxonomic classifications of reindeer and caribou, basically there are three different 'types': the continental or tundra group, which are highly sociable migratory herds; the more secretive boreal forest dwellers and mountain reindeer; and finally the diminutive high-Arctic survivors.

As a general rule, the larger the herd size, the further the animals have to migrate between their summer and winter range. The Barren Ground caribou of Arctic Canada are famous for the annual mass migration: from winter in the northern boreal forest, to traditional spring calving grounds on the tundra – a distance of 1,000km.

Broadly speaking, the migrating reindeer make up about 56 per cent, mountain reindeer 19 per cent, forest reindeer 14 per cent and high-arctic island reindeer 11 per cent. This current diversity came about through large-scale changes in distribution as continental glaciations advanced and retreated during the Pleistocene epoch, resulting in local adaptations of populations.

The females will travel an average of 50km a day to reach the calving grounds close to the Arctic Ocean. A short intense rut ensures all the calves (which will be many thousands) are born within a few weeks – a ploy to saturate the area with vulnerable young and so reduce the impact of predation. The status of these large migrating herds is relatively secure in terms of sheer numbers; however, certain herds have had their calving and grazing areas hugely compromised by industrial development and their traditional migration routes have been interrupted by enormous pipelines servicing the exploitation of oil and gas.

The North American woodland caribou are found in the northern boreal forest, which stretches from British Columbia to Newfoundland. This forest dweller tends to live close to or in the forest the whole year round, moves around in small groups and does not undertake vast seasonal movements typical of the tundra caribou. Sometimes known as 'the grey ghosts of the forest', woodland caribou can be very secretive and difficult to find. They generally live in small, scattered groups moving continuously through the forested areas. They are extremely shy. Females of the woodland-living caribou often have very small antlers, and a high proportion of the female Newfoundland caribou are actually antlerless. Because of their more solitary nature, woodland caribou do not need to compete for food as intensely in the wintertime and so it is felt that lack of competition means the females do not need to invest in growing antlers to be able to survive in the winter.

Another woodland dweller is the mountain caribou, which, by nature of their environment, make short annual excursions up into the mountains for the summer and travel back down for the winter. Their migrations are never usually longer than 50km, with an elevation change of 300m. These are some of the largest caribou in the world in terms of body size and antlers.

The high-Arctic reindeer are today composed of three island-living sub-species. The Svalbard reindeer is the smallest of all reindeer with a distinctive short face, short legs and a long, thick winter coat. Geographically isolated on the high-Arctic islands of Spitzbergen for some 20,000 years, today's population numbers about 10,000. These reindeer live in the most nutritionally demanding conditions and experience great fluctuations in climate throughout the year. They do not undergo any form of migration and they have no natural predator, hence the short legs. Overhunting by man during the early part of the twentieth century undoubtedly led to an uncertain future for the Svalbard reindeer. However, when the Norwegians took over the ownership of Spitzbergen they implemented a ban on hunting in 1925. This has secured the population for the foreseeable future.

Off the north coast of Russia in the Novaya Zemlya archipelago there is another distinct subspecies: Novozeml'sk reindeer. Like the Svalbard reindeer they have had a chequered history of overhunting but in recent years the population has stabilised at 10–15,000. Reindeer from these high-Arctic islands have an incredibly short growing season of about eight weeks and in this time the reindeer have to do all their maturing, growing of antlers and laying down of fat reserves. However, one bonus is the lack of mosquitoes, which torment the mainland species.

The Peary caribou is also small, though with longer legs than the Svalbard reindeer. They are restricted to the high-Arctic Queen Elizabeth Islands of Arctic Canada. Unfortunately, their population numbers have been on a downward spiral with an almost 70 per cent reduction between the

1970s and 1990s. Their decline has been attributed to many factors, including competition for grazing with another high-Arctic mammal, the musk ox; predation; hunting; and severe weather conditions. The Canadian government has now added the Peary caribou to its list of endangered animals and is actively trying to prevent their extinction. The population is currently estimated as 14,000, down from 22,000 in 1987.

The caribou of Alaska and Canada and the high-Arctic reindeer of Spitzbergen are truly wild animals, hunted by man but never herded or domesticated. However, as you cross the Bering Straits into Russia, it is a completely different story.

The Eurasian Tundra reindeer includes all reindeer, both wild and domesticated, to be found west of the Bering Straits, in the tundra and north taiga of northern Russia and Scandinavia. Domesticated reindeer far outnumber wild reindeer, which today are confined to areas of Russia, with an isolated population of wild reindeer in south Norway. By far the largest wild herd of Tundra reindeer

There is some debate as to whether the reindeer in south Norway are truly wild or in fact feral, i.e. domesticated reindeer that have reverted to the wild. Either way they certainly behave like a wild reindeer and are not herded but controlled by hunting.

in Russia is to be found in the Taimyr Peninsula, where the total is about 600,000.

The Eurasian Forest reindeer has a tremendous distribution throughout Russia, from Karelia in the west to the northern part of Sakhalin and southern Kamchatka in the east. They are mainly to be found in the Russian taiga zone, with some populations inhabiting the forests and mountainous areas of southern Siberia and north Mongolia. There is also a reintroduced herd of wild forest reindeer from Karelia in Russia into Finland. To cope with the considerable depth of snow in these forests during the winter, these reindeer have the longest legs and biggest feet. The largest reindeer in Russia are a woodland sub-species from Kamchatka in the far eastern portion of Siberia at the coast of Okhotsk. Overhunting has restricted its range to within the Kronotsky Nature Reserve.

The total number of wild Forest reindeer in the Russian taiga is difficult to estimate but is likely to be in the region of 60,000 and declining. Like the tundra reindeer, there is also a domesticated form of woodland reindeer.

◆ 3 ◆

SPRING

FOR ME THE start of the reindeer's year begins in spring, when the female reindeer, or cows, calve. Like other seasonal breeders, producing young in the spring means there is an abundance of newly growing leaves, grasses and shrubs to ensure that the lactating mothers and their dependant calves thrive.

Spring for most of us suggests better weather, maybe a few rain showers and longer days with a warming sun. But the Arctic spring can still be incredibly wintry with plenty of blizzards and snow still lying on the ground. To withstand these weather conditions, the young calves have a very high-quality coat when they are born. They lack the hollow guard hairs that adults have but make up for that with a very dense, soft coat.

Unlike most other deer species, reindeer calves are very precocious and on their feet following their mothers around within hours of birth. This is a strategy that makes them grow up quickly and avert danger and predation by fleeing rather than hiding. Weighing in at around 7kg at birth, reindeer calves are incredibly 'leggy' – rather like a foal – and their back legs are noticeably longer than their front legs, causing them to tip forwards, almost nose diving into the ground. Strong back legs are important for running fast and also clambering over obstacles: there are many holes and hummocks that can trip up a newborn calf.

During the calving season we like to have early contact with our newborn calves to check that they are healthy, strong and can reach the udder, but we also find this early encounter tends to set the precedent for a calf that grows up quite tame. In Inner Mongolia, the Evenki reindeer herders take this to an extreme whereby as soon as the calf is born they take it away for some hours, thus imprinting it on themselves before returning the calf to the mother to suckle. This reinforces the closeness these indigenous people have with their domesticated forest reindeer.

The behaviour of individual female reindeer during calving varies enormously among the Cairngorm herd, with many of the cows being totally accepting of our presence, some taking umbrage and battering us with their front feet, and others running a mile as soon as we approach. For us this just reflects the tremendous range of characters we have in the herd.

One of the most laid-back reindeer ever in our herd has to be Haze who not only allowed us to enjoy her newborn calf, but allowed other females who had lost their calves to take part in its raising. Often a cow who has lost her calf will try to steal another one. Many mothers would not tolerate such behaviour but I suspect Haze positively encouraged it, and over the years Shine and Gazelle have both shared motherhood with Haze.

Her strategy certainly pays dividends: having two mothers to get milk from means the calf grows very quickly – Caddis is one of Haze's calves that benefitted from extra milk and she's now one of the biggest and most productive breeding cows we have ever had in the herd.

Reindeer calves take a very small volume of milk (approximately 0.3l per day) but what they lose in quantity they gain in quality as reindeer milk is extremely rich in protein, fat and vitamins and has a considerable amount of dry matter. The cow's udder is small and well protected with long guard hairs surrounding the mammary gland, and with four small compartments of rich milk the udder is well protected from frosty conditions.

As the calves grow up and get wiser they see opportunities to steal milk. This is quite common in reindeer and easily achieved because reindeer mothers actually have four teats. This is only possible when the cow is suckling her own calf and then another one will sneak up and suckle at the same time. If the mother notices she will chase the intruder away but all too often she is looking the other way, intent on nursing her own calf and so doesn't realise what's happening. Quite often it is a young inexperienced mother with her first calf who gets picked on and older savvy calves creep in for a sneaky suckle while she is tending to her own.

Cows are generally sexually mature at 1½ years old and so will have their first calf when they are 2 years old. Sexual maturity, however, is determined by achieving a certain body weight (around 50kg) and if this is reached when they are 6 months old then cows will calve as a yearling. Following a seven-and-a-half-month gestation, a single calf is normally born. Twinning is extremely rare in reindeer and so it may seem strange that the reindeer cow's udder has four teats — especially if you compare this with sheep, who actually only have two teats, yet frequently have multiple births! In 1988, a young female Grayling produced twins that were slightly premature and sadly died. Twenty years

On 8 May 2018 one of our oldest females, Lulu, gave birth to twins. Both alive, one bigger than the other, and Lulu had managed to calve without any assistance. Indeed when reindeer herder Fiona found her, she was in a complete state of shock to discover there were two. Such a rare event, and I must say when I went onto the hill later and saw them myself it was a very special moment. There must be something in the number 8! In 1988 and 2008 sadly the twins born were dead, but now in 2018 we finally have live twins! From our research we have found only one other set of live reindeer twins who were born in Finland.

later Polo, a 10-year-old female, was found by our daughter Fiona struggling to calve. Fiona thought there was something up when there seemed to be an awful lot of feet. She managed to herd her back to a shelter pen and with veterinary assistance quickly ascertained that there was more to this calving than normal, as six legs were by then sticking out. Unfortunately the calving was difficult due to the two of them trying to exit together, which meant that both calves died. Poor old Polo, she had managed to get them to full term for nothing.

Caribou are one of the few deer species whose calves are unspotted at birth. They do not resemble the spotted coat of the typical 'Bambi' at all, as their normal coat colour is chestnut brown with a black back. Camouflage is vitally important for a vulnerable youngster and the combination of colours blends perfectly into the background of bare ground and patches of snow.

Through the year reindeer are generally quite silent except at calving time and early summer when cows and calves frequently 'grunt' to each other, particularly when the herd is on the move and they are trying to locate each other. The grunt of the calf is a bit like a duck quacking: high pitched and persistent. The cow responds with a low gruff grunt, which is repeated incessantly until her offspring is safely back at her side.

In 2008, we named one of our calves 'Grunter' because of his persistent grunting. To be fair to Grunter he was deserted by his mother Dixie at birth. When Grunter was born, Dixie took flight (or fright!) and despite attempts to reconcile them Dixie was having none of it, so we resorted to bottle-feeding Grunter ourselves. Having never been nursed by his reindeer mother, Grunter was inevitably pretty confused, and whenever he saw us, even as a mature reindeer, he would often grunt a hello!

I hasten to say we didn't grunt back, although when we are closely herding our reindeer, or encouraging them to follow us, we do call them with a low 'luh, luh' noise (almost a grunt). Which reminds me of an amusing story.

Some years ago I was up on the hill with fellow herder Colin, who had brought his friend from Lincoln up to see the reindeer. We were walking along a low ridge chatting to each other but simultaneously I was encouraging the reindeer to follow us across the hillside. Every so often I would interrupt our chat with a 'luh, luh' to make sure the reindeer were still following. Later that day Colin's friend Warren commented on what a great afternoon he had had with the reindeer but he said, 'Shame about Tilly's speech impediment'!

4

MR UTSI

IN THE SPRING of 1952 a small group of domesticated reindeer landed at Rothesay dock, Clydebank, en route to their new home in the Highlands of Scotland. These reindeer belonged to Swedish Sami Mikel Utsi and were the beginning of the herd that has now been present in the Cairngorm Mountains for the last 65 years.

In the late 1940s a proposal had been mooted to Britain's Ministry of Agriculture to reintroduce reindeer to Scotland. Dr Lindgren, a social anthropologist, and her husband, Mikel Utsi, were the driving force behind this suggestion. Mr Utsi was an expert reindeer herder and Dr Lindgren had all the diplomatic skills required to woo the authorities in the British government. Better still, this formidable team was privately financed. Having gained approval for their plans, they searched for the best place in the Scottish Highlands to bring the reindeer to, and then devoted the latter halves of their lives to the project's success.

Married late in life, Mr Utsi and Dr Lindgren were an extraordinary couple. Mikel Utsi was Sami, spoke only broken English and stood just 5ft 4in tall. Born in Karasuando in 1908, Utsi was the second of eight sons of a well-known reindeer-herding family. He grew up in a traditional nomadic life, living and breathing reindeer lore as the family followed their reindeer on their seasonal migrations. Karasuando is close to the Norwegian border and the Utsis' reindeer would migrate into Norway for the summer, returning to Sweden for the winter. But in the early 1920s this was stopped because the Norwegian and Swedish governments agreed to a 'hard border', which put an end to reindeer movements across it. The Utsis, along with other reindeer-herding families, were forced to move further south and re-settle, wintering their reindeer in the forests around Jokkmokk and herding them up to the Swedish mountains, along the Lulea river for the summer. During the late 1930s and the Second World War years the resourceful Mikel Utsi had a varied career. He ran restaurants, served in the Swedish army and as a Swedish special constable helped rescue many hundreds of Norwegian refugees escaping from the Germans across the wild, mountainous borderland. After the war he was awarded the Freedom Medal by Norway's King Haakon.

Dr Ethel John Lindgren was the daughter of a Swedish–American banker, she had studied Chinese and anthropology at Cambridge University, spoke at least five languages fluently, had taken part in various ethnological expeditions to indigenous people as far afield as Outer Mongolia during the late 1920s and early 1930s and stood over 6ft tall. The couple met while Dr Lindgren was in Swedish Lapland studying the Sami people. Perhaps she felt her only chance of persuading Mr Utsi to get married and come and live with her in Britain was to promise that he could bring his herd of reindeer with him!

When Mr Utsi visited the Scottish Highlands at the end of the 1940s he was struck by the similarities between the Scottish landscape and habitat and his own reindeer-grazing lands in north Sweden. He was convinced, quite rightly as it turned out, that reindeer could do well in this new land, especially since they had lived here perhaps as recently as 800 years ago, and Scotland offered a habitat very similar to their homeland Lapland. Another part of the argument for the project played on the fact that food rationing was still in place in Britain after the Second World War and the reindeer would provide a very welcome source of meat.

In The Orkneyinga Saga it is written that the Vikings hunted reindeer and red deer in Caithness and Sutherland. Certainly they would have known the difference between the two species as both are native to Scandinavia. However, there is no archaeological evidence to substantiate this claim. Perhaps the Vikings themselves brought the reindeer across – a not unlikely suggestion since reindeer were certainly domesticated by the Viking era. The last evidence for wild reindeer in Great Britain dates to what was probably an interglacial phase some 3,000 or 4,000 years ago.

So it was that in 1952 the first small consignment of eight reindeer was taken by boat from

Sweden via Narvik in Norway to Scotland. There were two bulls, one castrated male (ox), two young females and three female calves. They arrived on 12 April 1952, and were transported immediately into quarantine for twenty-eight days at the Royal Zoological Society in Edinburgh. Sarek the ox seemed rather an unnecessary addition to a breeding group, but Mr Utsi was hoping to do more than just bring reindeer back to Scotland – he was also bringing his Sami culture, in which castrated males play an important part. In fact Sarek, with his bell round his neck, was the herd leader and the pack animal, and if Mr Utsi hadn't been sentimental then Sarek would have been the first reindeer to be eaten. Instead Sarek went on to lead the herd for many years, dying eventually at the grand old age of 16 years.

Subsequent consignments were brought in over the next couple of years, with the third consignment causing a considerable stir among the authorities when Mr Utsi decided to bring one more animal than had been permitted, in his words 'to cover any losses that may occur en route'. It took all Dr Lindgren's diplomatic skills to smooth the waves following a formal letter of protest from the Ministry of Agriculture at Tolworth.

The reindeer brought in during the early years were kept on low ground on the Rothiemurchus Estate, beside Aviemore. The late Colonel Grant,

intrigued by the project, provided them with a few hundred acres of land and a cottage at Moormore for the herder. There were problems with the low ground and without a doubt from the very start Mr Utsi had his eye on the Cairngorm Mountains that dominated the landscape to the south of Moormore. By 1954, Dr Lindgren and Mr Utsi had persuaded the Forestry Commission for Scotland to let them move their modest herd of reindeer up on to the northern slopes of the Cairngorms. Some 2,428 hectares (6,000 acres) of open moorland, rising to 1,200m (4,000ft), were soon made available to the reindeer. This move to higher ground, where the cool breeze kept the flies away in the summer and the vegetation all year round was more appropriate, was a turning point for the herd and really set the project on a new footing.

The reindeer inevitably attracted a lot of attention from various quarters. The general public, amused by the prospect of Santa's reindeer being resident in Britain, were made welcome and from very early on Mr Utsi proudly showed visitors his beloved reindeer. Often dressed in traditional Sami costume, he caused quite a stir and many people remembered the charismatic Mr Utsi more than the reindeer themselves. A visit in the spring of 1953 from an ill-informed SSPCA officer resulted in a complaint being sent to the Department of

Agriculture, alleging neglect of the reindeer. What the officer didn't appreciate was that he had visited the reindeer when they were in full moult. The moulting of the light-coloured winter coat contrasting with the short dark summer coat underneath make even reindeer in the finest condition look particularly moth-eaten. A subsequent report by the Department of Agriculture completely vindicated Mr Utsi, but it made him think twice about taking visitors to see the reindeer when they were moulting.

In the hope of finding cheap meat supplies for his company's tinned dog foods, a representative from Chappie Ltd visited the herd and offered Mr Utsi 6*d* per pound for any surplus reindeer meat. Certainly one of the primary aims of establishing reindeer in Scotland was to provide meat, but not dog meat, and in Mr Utsi's words the Chappie representative was 'shown the door'.

The early years on virgin ground were challenging times for Mr Utsi, and the herd grew slowly. By March 1956, the herd total was twelve and the subsequent calving produced another four calves. Early losses through accidents, dog worrying and straying were devastating at the time, but they merely set the project back. Mr Utsi's zeal and determination kept the experiment on track and by the end of the 1950s the herd was going from strength to strength.

Although Mr Utsi spent a lot of time in Scotland with the reindeer he did not live there permanently and in his absence he depended on assistants to care for the reindeer. His expectations were very high and he demanded 110 per cent effort from all his volunteers and employees. He was certainly a difficult man to please and many people felt unable to keep up with his demands. On one particular occasion two volunteers were out looking for stray reindeer and they accidentally let go of their two reindeer, still with their halters on and ropes trailing. Mr Utsi was furious and sent them back out to find the reindeer with the trailing ropes. When they returned without his precious reindeer, Mr Utsi drove them straight to the station in Aviemore and put them on the next train south.

The original reindeer brought in from Sweden had a tendency to stray. They had come from migrating herds of reindeer and so had a strong urge to move with the seasons. Much of the work involved gathering up these strays from all around the Cairngorms and bringing them back to base. Neighbouring sporting estates, conscious of the foreigners in the hills, were regularly asked to look out for strays. One gamekeeper, approached to see if he had seen any reindeer, asked if there had been any interbreeding with red deer, 'because I have seen a deer with a very odd-shaped pair of antlers'. There were also reports of 'ghost reindeer' in Glen

Feshie on the west side of the Cairngorm range, but Mr Utsi put this down to 'a few too many tots of whisky'.

Volunteer reindeer herders came from all walks of life, from locals to university students. Their accommodation moved from Moormore Cottage on the Rothiemurchus Estate to Reindeer House, a substantial bungalow built by Dr Lindgren and Mr Utsi in Glenmore in 1960. Commanding a fine position in the Glenmore Forest and with a clear view of the northern slopes of the Cairngorms, Reindeer House was ideally positioned for the keepers to watch the reindeer on the hill. Reindeer House was very much regarded as a place of work rather than a home, and Mr Utsi was very specific about the fixtures, fittings and general decor. There were no comfortable sofas or settees and the only source of heat was a small coal fire. Apparently he believed that if he provided the workers with too much comfort they would become lazy and stay in the house. It is perhaps not surprising that over a period of thirty years Mr Utsi saw more than 500 staff, both paid and voluntary, come and go.

By the mid 1970s Mr Utsi's health was failing and he was no longer able to walk up into the hills to see his beloved reindeer. Frustrated by illness, he had to depend more and more on his assistants to look after the herd. The hot summer of 1976 took its toll on the herd, but Mr Utsi, depressed by the losses, blamed the keepers instead. His journeys north to Glenmore became less frequent and the demoralised workforce did their best to keep tabs on the reindeer. By the autumn of 1978 Dr Lindgren was once again looking for another keeper to tend the herd on the hill.

5

SUMMER

ALTHOUGH THE WINTER months are long dark days, the only light coming from moonlit snowscapes, the exact opposite occurs in the Arctic summer. From the spring equinox the days lengthen rapidly until midsummer when the sun never sets below the horizon and there is daylight 24 hours a day. This brings an explosion of life among the Arctic plants and it is this abundance that enables reindeer and caribou to thrive in the north.

Over the summer months it is essential for the reindeer to lay down enough fat reserves to see them through the lean times in the winter. There will be many occasions when the snow is too deep or too hard to dig through and the reindeer will have to go without food of any sort. Sufficient body reserves are needed to bring them through these periods of starvation. Despite all their adaptations to survive the cold, without the profusion of growth in the Arctic summer reindeer and caribou would never be able to prepare themselves sufficiently for eight months of snow. They are supremely adapted for their winter environment – but they are equally well adapted to make the most of the summer's bounty.

It is imperative that the calf grows quickly from birth. Nutritious reindeer milk and good summer grazing means that a calf can double its birth weight in two to three weeks, and reindeer calves may weigh as much as 50kg by the autumn. In addition to this rapid growth, and laying down fat reserves for the winter, the calves also have to grow their first set of antlers, moult their original calf coat and grow their first summer coat, followed in quick succession by the onset of growing their thick insulating winter coat. This all has to happen in a comparatively short space of time during the summer and autumn to ensure survival through the winter months.

In its first few weeks of life the calf is never far from its mother's side, lying close to where she grazes, always following her when she moves on to fresh ground. As the calf grows older it becomes more adventurous, leaving its mother's side and spending more time with the other calves. Forming 'nurseries', the calves enjoy one another's company and often become playful in the twilight hours. From small protrusions on its head, the first set of antlers begins to grow when the calf is around a month old. Over the next few months the antlers will grow in order to enable the calf to compete for food in the winter months. The size of the calves' antlers by the end of the summer will vary greatly depending on the quality of their diet. A fully grown set of calf antlers may be as long as 40cm with a couple of points. By the end of the summer, the calves become 'mini adults', with all the wherewithal to face the rigours of winter.

Caribou and reindeer are described as mixed feeders, browsing and grazing fastidiously on fresh and untouched green vegetation. Some plants they consume completely, others only partially. They eat 'on the hoof', never staying in one place for long, and where they can they make the most of the advancing spring as they move north towards the Arctic Ocean or up into the high mountains.

The green plants contain all the nutrients and vitamins required for growth. Willow and birch shrubs are very valuable fodder, being high in protein and minerals. Leaves are plucked and eaten as soon as they appear, and continue to be consumed avidly through the whole of the summer. Sedges and cotton grasses before they flower are also rich in protein and are eagerly grazed in the spring when they are juicy and tender. Cotton grasses are extremely important in some areas, representing as much as 90 per cent of the summer food. The primitive horsetails are also enthusiastically eaten.

Herbaceous plants are also browsed as they too are rich in nutrients. They tend to develop later than grasses and sedges and are a very important component of the reindeer's diet in late summer before the first frosts turn them brown and unpalatable. Lady's mantle, marsh marigolds, cloudberry, lousewort, sorrel, bogbean and irises are all consumed during the summer months.

The end of the summer also sees the fruiting bodies of mushrooms, which are a highly prized source of food for reindeer and caribou. Although mushrooms consist of over 90 per cent water, they are rich in proteins, have a high potassium and phosphorus content and contain numerous vitamins. Reindeer can detect the scent of suitable mushrooms from some distance, and their scattered distribution will cause the animals to move about erratically in late summer. Highly favoured are the boletus mushrooms, which are eagerly devoured.

The behaviour of our reindeer is extremely erratic at this time of year when they are on the quest for mushrooms. They will often range well into the forest margin looking for them and have even been known to sample the wrong type, which can cause them to have sore stomachs and on occasion to hallucinate. Two of our Cairngorm reindeer, Frostie and Bourbon, have both suffered from indulging in the wrong mushrooms. Frostie was the funniest because he appeared to actually hallucinate and came down the hill basically walking sideways! He must have enjoyed it because the next summer he did exactly the same thing. Bourbon we assume ate a different type of fungi because he just suffered from a sore stomach. In his case we were certain that fungi had caused it because he had a radio collar on and we were able to pinpoint exactly where he had been. There was even a bite out of the mushroom!

Reindeer and caribou are thought to eat more than 350 different species of plants during the summer months in their quest to prepare themselves for the imminent falls of snow, plummeting temperatures and frozen ground.

This feast of food in the summertime is also what makes it possible for reindeer to grow new antlers annually – a process that requires lots of high-quality vegetation. The mature bulls grow the biggest, most complex antlers. Although there is a basic format to reindeer bull antlers, which make them distinguishable from the antlers of other deer species, there are considerable differences between sub-species and indeed individuals within any particular herd. Individual reindeer antlers have a unique pattern that is embellished each year, becoming larger and more complex each time until the bull reaches his prime.

For us, Crann has been the most impressive bull we have bred and at the age of 8 years old grew the largest antlers ever in our herd. He died an old reindeer in the summer of 2017 at the grand old age of 14 years. I suspect it will be a long time before we have another bull of his calibre. So why did Crann grow such big antlers? Well, it is probably a mixture of nature and nurture. His mother Burgundy always grew big antlers for a female; she also didn't have her first calf until quite late in life so she was always in fine condition and had plenty of milk for her calves. Crann also grew up to be an extremely greedy reindeer, always there for extra hand-feeding and never pushed around by young upstarts. His prime years as a rutting bull coincided with us importing new reindeer from Sweden to improve the genetics of the herd. This meant that Crann was never used to his full potential as a breeding bull and so always began the winter in good condition. Normally, bull reindeer start growing their new antlers about March but Crann was always about six weeks earlier, often showing considerable length even in January. So by the middle of August when the antlers ceased growing and the velvet peeled away Crann had by far the biggest antlers. He was in a class of his own.

There is a 'predator' that dominates the summer grazings, and although it doesn't kill its prey, it drives them to distraction. During the summer the Arctic regions are awash with biting insects, from midges in the Scottish Highlands to mosquitoes in the Arctic tundra, and they all feed on the blood of mammals to survive and breed. The female mosquito needs blood to lay viable eggs and so focuses on the areas that are less hairy or have thinner skin. The newly growing velvet antlers are particularly vulnerable and on warm

sunny days these insects can make life pretty miserable for a caribou. By forming large groups the caribou try to reduce the level of harassment, presumably by passing the problem on to their neighbours. Coastal breezes and permanent snow-fields both help to ease the effect but the insects are savage and the caribou find themselves in a catch-22 situation, where food is abundant but the insects prevent efficient feeding.

Luckily here in Scotland we don't have mosquitoes but we do suffer from midges, which are equally attracted to the thin skin of the velvet antlers. On hot humid days in the summer, clouds of midges and head-flies (they're not called *Hydrotaea irritans* for nothing!) make the reindeer very agitated – they stamp their feet, run around in circles and try to find a breeze up on the ridges to blow the irritating insects away.

Reindeer only moult once a year, but it is such a thick dense winter coat to lose that it takes many weeks before the short dark summer coat is completely revealed. Reindeer can look pretty scruffy during moulting, with the first signs of this starting round the eyes before spreading across the body. The old coat comes off in handfuls and when they shake it is like a 'snowstorm of hair'. Like the short summer season the reindeer are actually in full summer coat for a relatively short time and within a few weeks of moulting their old winter coat, the beginnings of the new pale-coloured winter coat begins to grow first round the neck and shoulders, contrasting with the darker summer coat of the body.

The first autumn frost puts an end to the tormenting insects but it also heralds a change in the vegetation, which begins to turn brown, losing its high nutritional value. By now the reindeer need to be well endowed with layers of fat and thickened coats. I often liken the autumn colours of the vegetation with the changing colour of our reindeer. Both the vegetation and the reindeer need to prepare in good time for the winter – it is never far away once autumn has arrived.

• 6 •

ETHEL LINDGREN AND HER 'CROCODILE DUNDEE'

WHEN PEOPLE RECALL the history of the Cairngorm reindeer they invariably think of Mr Utsi as the person who devoted his life to the success of the project. Mr Utsi was certainly the 'front man', always keen to introduce visitors to his beloved reindeer and any newspaper or magazine articles had photos of him, dressed in his Sami costume with reindeer. Indeed visitors to the reindeer in the early days remembered Mr Utsi more than the reindeer. He was sociable, enjoyed many a dram with the locals and was a bit of a one for the ladies too! He loved nothing more than showing off his reindeer, on the hill, at agricultural shows and Highland Games.

But behind the scenes his wife, Dr Lindgren, was a force to be reckoned with, and without her diplomatic skills, financial backing and determination the project would never have even got off the ground. Her early life was extraordinary and as a very private person she never 'blew her own trumpet'; indeed she maintained an element of secrecy about her past. Until she passed away in 1988 I knew very little of her life so it was a revelation to me when I heard the whispers were true.

Born in Evanston, Illinois, USA, on 1 January 1905, Ethel John Lindgren was the only child of a wealthy Swedish–American banking family. She travelled extensively with her family: at the age of 15 she went to Japan, at 16 she did the grand cultural tour of Europe and as a 17-year-old she rode north on horseback to view the inner sections of the Great Wall of China at Kalgan, where she was impressed by the 'vast open landscape'.

At the age of 19 she was awarded a college scholarship to attend Newnham College, University of Cambridge, to read Oriental languages and Moral Science. Three years later she had completed her degree with first-class honours. She was multilingual, speaking Swedish, French, German, some Dutch, Mongol, Tungus and Russian (informally). As a postgraduate she went to Ulaanbaatar for field research but political unrest in the People's Republic of Mongolia did not permit her to carry on her studies successfully and she was virtually under house arrest for some months.

It was here that the young Lindgren teamed up with Mongolian veteran and Norwegian national, Oscar Mamen. Together they left Ulaanbaatar, arriving in the town of Hailar, China, to try to pick up the pieces and find another area of ethnographic research. It was here that they heard about 'a far-off tribe of reindeer herding people who were fast dying out'. Known then as the Reindeer Tungus, these people were thought to originate from Russia, crossing the border into China where they settled between the western foothills of the Kinghan Mountains and the east bank of the great River Argun that forms part of the border between

Russia and China. Living in this extremely remote area with many swift-flowing rivers, the Reindeer Tungus' only contact with other people was when they visited Russian émigré and Cossack outposts. Here they would bring skins of animals they hunted (squirrels and martens) and barter for provisions such as flour, cloth, tobacco and spirits, tea and sweets.

Ethel Lindgren and Oscar Mamen set off with Haisan Gobul, a local man from Hailar, travelling north along the River Argun, before heading east along the fast flowing tributaries where they hoped to meet up with these people. At Chuerhkanho, a Cossack outpost, they persuaded a local Russian Cossack to work as a scout in the dense taiga where these reindeer people lived. It was not easy to persuade him, as the Cossacks seldom entered these areas during the summer because of the many rivers, marshy ground and numerous biting insects. The only time they normally ventured here was in the winter, when the rivers were frozen and the insects absent.

As luck would have it, smoke from their campfire was spotted by a Reindeer Tungus who was out hunting. Following a meal of 'meat and vodka', the man was willing to lead them back to his camp. Their first sight of the reindeer camp with 'tepees' made of alternating white and brown birch bark in the middle of birch and evergreens

was very rewarding and Haisan was fascinated by the 'tame deer' whose existence he had frankly disbelieved. During the day the reindeer usually lingered around the camp and the click of 'little hooves' attached to salt bags was enough to bring them running to the tepee where they were wanted. It was midsummer and the womenfolk would milk the reindeer three times a day – while small in quantity, it was delicious and as thick as cream. During the night the reindeer would roam far from the camp in search of food, eating not only lichen but the leaves of birch, aspen and various small green plants too.

Summer was a time of the year when the Tungus broke camp less often. The men would go off in advance to blaze the trail, cut poles for new tepees and start fires. The women who were left behind would do all the packing, loading and guiding of the reindeer. Maddened by mosquitoes, the Tungus only travelled at dusk, so the reindeer stood quietly enough to be loaded, and they were tied in small groups beside smoky fires. A heavy smoke was made by heaping on Labrador tea, a fragrant shrub, which gave off a strong pungent odour, deterring the annoying insects.

On this first and most cumbersome journey it was Oscar Mamen, fluent in Mongolian, resourceful and extremely capable, who took on the role of main provider and official photographer. In one of many letters home to her mother, Ethel praised Oscar Mamen 'for being all and more than he is advertised to be, with plenty of Norwegian gloom when things are going well, but determined optimism when things are really bad, which on this trip ... was often enough'. From reading between the lines I think he was Ethel's 'Crocodile Dundee' and without him she would never have ventured into such remote territory.

The two of them left China in December 1929, to return to Cambridge (for Lindgren to give lectures about their trip) and via Lillehammer in Norway, where they married. They then went to Lindgren's family home in the USA and at the end of the summer Ethel Lindgren gave birth to her only child, John. When John, still a babe in arms, was just 2 months old his parents left, hotfooting it back to China to continue their research for the next two years.

And so their second trip to the Reindeer Tungus was not until October 1931 when they used the knowledge of their previous expedition and chose a more direct route through the taiga to the upper reaches of the River Bystraya, arriving in November. Here they lived with these nomads for twenty days.

On this second trip Ethel Lindgren established a close friendship with Shamaness Olga Kudrin Buldotov. Olga was able to provide Lindgren

The word 'shaman' comes from the Evenk hunters and reindeer herders in Siberia. However, by the twentieth century, it became widely applied in North America to a wide range of medicine men and women. The Siberian shaman's soul is said to be able to leave the body and travel to other parts of the cosmos, particularly to an upper world in the sky and lower world underground.

with rich details of these people's lives and Lindgren was particularly interested in the role of women – their status and daily chores – and shamanism. In these early days of ethnographic studies, many of the researchers were men who focused more on the male culture and so Ethel Lindgren's emphasis on the womenfolk was a refreshing change from the male-dominated view of life.

Every few days the nomads would pack up their belongings and move to areas of fresh grazing, the whole process documented in detail by Lindgren, who took copious field notes, and Mamen, who took hundreds of still photographs and cine film. Indeed there are two fascinating short films, edited with text, like a 'Chaplin film' of the lives of these people.

Although there was plenty of lichen for the reindeer to eat, there was nothing digestible for the horses the couple had arrived on and their visit was brought to a sudden halt by the critical condition of the horses. In spite of forced marches back to where oats awaited them, three horses died on the trail.

The third and last journey to these people was from May to July 1932. Again they met up with Olga the shamaness and gleaned more knowledge, photographs, hunting implements, household items, panniers for packing the reindeer, saddles, cradles and clothing. But by now Manchuria had been invaded by the Japanese and the puppet state of Manzhouguo was founded. Anthropological fieldwork was no longer possible and the couple returned to England where Ethel Lindgren wrote her PhD thesis, 'The Reindeer Tungus of North West Manchuria', a study of Tungus society, including an interpretation of a shaman's activities and function in the community.

For Lindgren and Mamen an English lifestyle and academic career at Cambridge University took its toll on their marriage and it was not long before the two of them parted company. She then turned her attention to another reindeer herding tribe and headed for Jokkmokk in North Sweden where she would meet her next 'husband to be', Mikel Utsi.

• 7 •

AUTUMN

B Y THE END of the summer reindeer are look-ing at their very best, with full sets of antlers, layers of fat for winter and thickening winter coats. It's time for the rutting season.

With a gestation period of seven and a half months, conception needs to take place in the autumn to ensure the calves are born in spring. The rutting season is a time of strenuous activity for the bulls, who have spent the summer prepar-ing themselves.

Bulls normally breed for the first time at 3 or 4 years old and will usually reach their prime at about 6 years old. The contest between reindeer bulls entails both an element of display and a show of strength, although a degree of experience will also stand a bull in good stead. During the weeks leading up to the rut the neck muscles of a mature bull swell and this combines with increased body-weight to form the powerhouse with which they will do battle. The antlers, now stripped of their velvet, act both as ornaments to display with and as handles to lock the heads of two bulls together for a pushing match. Displaying between two bulls often involves parallel walking, where the bulls will walk side by side, eyeing each other up. In some instances the sheer size of a bull's antlers is enough to deter a prospective challenger. However, if the display alone is insufficient, a fight will commence. The bulls lock antlers and push. During this battle

of strength they may break their antler lock and re-engage a number of times, turning on a sixpence before clashing heads again. Battles may last for some time. We have first-hand experience of bulls fighting for dominance most years in our herd and if the bulls are evenly matched the fight can go on for a considerable length of time.

In 2014, two of our own reindeer bulls, Bovril and Gandi, had one such encounter. Bovril was already holding a group of cows and Gandi arrived, looking to take over. It was fascinating to watch as Gandi was one minute strolling along, taking in the scenery and suddenly spotted Bovril, further up the hill with his group of females. Nothing happened quickly as they both assessed the situation and Bovril waited for Gandi to approach him – Bovril had everything to lose. They eyed each other up, lowered their heads and finally locked antlers and did battle. The hillside has plenty of obstacles to trip them up and the bull on the upside always has the advantage, so there was plenty of whirling around, re-connecting of antlers and sustained pushing. The minutes ticked by as the two bulls remained in constant combat, with no sign of either letting up. They were just so well matched and there seemed to be no end to their ability to keep fighting so strenuously. After an hour of constant battle and with no obvious end in sight we decided to intervene. Luckily their battleground had moved and they were inching closer to a gateway. So when the opportunity arose we managed to open the gate, encourage Bovril's cows to go through and he took the opportunity to break from fighting and slip through with them. Don't worry, we made sure Gandi got a group of cows too!

Although injuries do occur, the outcome is seldom fatal. The winner of any contest will drive the loser away and will remain dominant over his opponent until such time as he tires, breaks his antlers or becomes injured.

As well as chasing other prospective bulls away, the dominant bull must constantly gather his wandering females (they don't make it easy for him) and of course copulate with any female that is receptive. The breeding bulls are so busy they have not the time or inclination to eat and so rapidly lose weight. This creates a potentially dangerous situation as they then have to face the rigours of the winter in poor shape. Indeed there can be substantial mortality among the breeding bulls during the winter months, which is a contributory factor in average life expectancy. Bulls tend to have a shorter life expectancy than the cows.

While the bulls are frantically trying to pass their genes on to the next generation, the cows are conserving their energies for the winter. Their

udders dry up so milk stops being produced and the calves are naturally weaned. Like the bulls, the cows and calves strip the velvet from the newly grown antlers and now have these 'weapons' ready for doing battle in the winter.

It is generally agreed that the females and calves grow antlers so that they can successfully compete for food in the winter. The mature bulls are the first to lose their antlers, as soon as the rut has finished, and they immediately lose status, falling to the bottom of the pecking order, even below calves with their small antler sticks. The cows have antlers of a similar size to the young bulls and so can compete favourably with them when it comes to finding food. As the winter progresses the young bulls cast their antlers, leaving the cows to rule supreme for the rest of the winter. They will not cast their antlers until they calve in the spring.

> Antlers are a status symbol and without them a reindeer falls down the pecking order. So losing them is both stressful for reindeer mentally, because they can be bullied by others who still have their antlers, and physically because when the antlers breaks away from the skull there is an exposed bloody wound.

With autumn sliding into winter the cows will carry a growing foetus through the winter months.

The mother must make the most of the snow craters she digs to find food as her calf of the preceding spring will benefit from her status in the herd. There will be a strong bond between the two all winter – a bond that, between mothers and daughters, often lasts for a lifetime.

8

THE SMITHS

IT WAS OCTOBER 1978 and farmer's son and recent school leaver Alan Smith was looking for a job. His mother Bunty, an avid reader of the *Press and Journal*, found an advert in the jobs section – 'Reindeer Herder wanted' – and she encouraged Alan to apply. Technically Alan did leave school, but to be absolutely honest he was never really 'at school', being far more interested in the great outdoors, helping on the farm, deer stalking and grouse beating. On the one occasion when he did turn up at school to take his exams the teacher recommended his time would be better spent going back to what he was doing – 'cutting peats' for fuel during the winter.

As luck would have it Alan gave as his referee Commander Carmichael, who owned the neighbouring estate to the one his father worked on. Alan had done some deer stalking for him, and it turned out that the commander knew Dr Lindgren and Mr Utsi very well. In fact, he was a member of The Reindeer Council of the UK, a research body Dr Lindgren had created when the reindeer were first brought into Scotland in the 1950s. The old-boy network worked a treat and Alan got the job. Alan moved into Reindeer House to start his new job on 23 November 1978. He was 18 years old, had never seen a reindeer before and had no idea what he was supposed to do. Brought up on a hill farm, I think he imagined himself driving a tractor and trailer and feeding the beasts hay and silage. He could not have been further from the truth! In fact, his back yard was the Cairngorm massif, Britain's largest sub-Arctic plateau area, and the only means of access was on foot. Luckily for Mr Utsi, Alan was quite unfazed by the prospect of tramping the hills looking for reindeer. Within days of arriving he spotted some reindeer about 4 miles away, grazing on the far skyline. He phoned Mr Utsi in Cambridge with the news that he had located part of the herd, only to be told, 'You must have a good pair of binoculars.' Mr Utsi never made it north in the last year of his life and so Alan was very much left to his own devices. Although daily diaries had been kept, there was no hands-on expertise to be passed down and from day one Alan had to learn the ropes the hard way. Mr Utsi died the following June in Cambridge and with his death a tremendous amount of knowledge was lost forever. Alan was Mr Utsi's last reindeer keeper and with his passing Alan was left 'holding the baby'. Dr Lindgren was still in charge of the Reindeer Company but she had no practical skills to hand on to Alan.

What Dr Lindgren did have though was contacts and she was constantly providing Alan with volunteers. I suspect this was more to keep an eye on Alan and report back to her rather than do anything useful. Although Dr Lindgren lived in Cambridge she regularly drove north to stay at Reindeer House. This is no mean feat for anybody and she continued to do this four or five times a year until she died at Reindeer House at the age of 83. In her later years she was never fit enough to go out onto the hill where the reindeer graze but she spent all her waking moments at Reindeer House, socialising with locals and meeting with influential people. Her daily contact with Alan was by phone while in Cambridge, but daily interrogations when at Reindeer House. She always liked to speak to Alan or his assistant alone, never at the same time. She was also a serial letter writer. Every letter she posted to Alan was almost indecipherable as her handwriting was atrocious. And at the top of most of her letters it read 'Private and confidential, destroy immediately once read'. Needless to say many a letter was left unread by Alan. She also wanted to know everyone's business and very naughtily opened many a letter addressed personally to Alan! After opening it she would scribble across the envelope, 'opened in error', with not so much as an apology! What kept Dr Lindgren the happiest was if Alan actually got round to writing and posting a daily diary. But this was not his forte and the contents of many of Dr Lindgren's letters to him were about the need to keep and post to Cambridge daily records.

I first met Dr Lindgren as a teenager, when she came to my father's local Natural History Group to give the 'Christmas talk' on reindeer. I remember the front door of our home opening and this larger-than-life figure, Dr Lindgren, sweeping in, in full evening dress. Even my father was a trifle overawed. He made suitable welcoming noises and hoped the bad weather would not put people off turning out for the talk. 'I do hope not,' she replied sternly, clearly implying that a little wintry weather in the south of England was nothing compared with the Arctic winters she and her reindeer were used to.

Some years later when I was finishing the final year of my zoology degree I decided to write to Dr Lindgren to ask if I could work as a volunteer. On receipt of my letter she summoned me to meet her at her home near Cambridge. It was a very daunting experience. The gloomy interior of the house and her interrogating manner was certainly memor-able but I must have impressed her enough (or she must have been desperate) because by the end of the afternoon she announced that I could assist the reindeer keeper as a volunteer for the summer. I got the impression that she imagined my presence at Reindeer House was going to be more one of reporting back to Cambridge on the daily goings-on at Reindeer House, rather than usefully helping the reindeer keeper. Little did Dr Lindgren,

or I come to that, know that actually she was sending me north to meet my future husband.

As I drove north on the 500-mile journey to Aviemore I imagined Reindeer House to be a large rambling shooting lodge, perched on some remote hillside. My thoughts could not have been further from the truth. Reindeer House is a relatively modern bungalow, built in 1960 and nestled in the midst of the small Forestry Commission community of Glenmore, across the road from a thriving campsite. Having driven almost the length of Great Britain I was mortified to find no welcoming party; indeed the house was empty with no sign of human activity at all. I wandered round the back and asked a neighbour as to the whereabouts of the reindeer keeper. 'Oh, he'll be away fishing,' came the reply. 'But you can let yourself in; the door's always open.'

Sure enough, the door was not locked and I let myself in. I felt a little like Goldilocks entering the bears' cottage! From a room there was frantic yapping, which turned out to be a highly delighted, very small and very hairy dog, who proceeded to follow me around the house as I investigated. In need of some reassurance, I decided to ring home and tell Mum I had arrived. But where was the phone? I eventually found it: a pay phone in a cupboard. But while I was ringing home, the keeper, Alan, was ringing Reindeer

House to see if I had arrived! On hearing the engaged tone he immediately assumed I was on the phone to Dr Lindgren, reporting his absence. Dr Lindgren had given Alan the impression that I was much older than him, extremely capable and built like a tank. I think he was rather dreading his first encounter with me. It transpired that he had not gone fishing at all but had gone to Aberdeen for the day with friends. When he eventually returned and I introduced myself he eyed me in a very guarded manner. But I was not an 'old battle axe' as Dr Lindgren had led him to believe and I think he was quite pleasantly surprised by what he saw!

I suspect as a little tester to see what I was like, Alan announced that two reindeer, Eidart and her calf, who were in the paddock beside the house, needed to be led up onto the hill. This was my very first encounter with a real live reindeer and I was immediately struck by how small they were. It was a beautiful June evening and Eidart and her young calf were eager to get back to the hills. With Eidart on a halter her calf willingly trotted along beside her. The 2 mile walk up to the hill to the rest of the herd was my first indoctrination into reindeer herding. And if there was more of this hands-on work with such a gentle, willing creature then working with reindeer in the Highlands of Scotland was definitely a yes from me.

Alan and I got on famously. I helped him look for stray reindeer on the hill, kept up his daily diary and took visitors to see the herd. The combination of breathtaking scenery, endearing reindeer and a Scotsman who wasn't bad looking soon persuaded me that my future lay with the reindeer and their keeper! So when autumn came I decided to stay, much to Dr Lindgren's considerable discomfort. I was instantly 'persona non grata', a spanner in the works and definitely not 'the mole in the camp' she had hoped for. She even suggested to my parents that they should 'remove their daughter from the north'. In the meantime Alan and I had a whale of a time. Life carried on very much as normal, except I became an unofficial resident. The neighbours' tongues were wagging, no doubt having had their ears bent by Dr Lindgren, and when official visitors came in through the front door, I would make a hasty retreat out of the window. But it was all quite harmless and I was unpaid help to Alan!

All this French farce eventually came to an end when, much to Dr Lindgren's relief, Alan and I got married. The 'Keeper's room' became our sitting room and the guest room our married quarters. I immediately became acceptable again to Dr Lindgren, and Alan and I set about looking after the reindeer as an official duo. For the next five years Dr Lindgren would make regular journeys north to stay at Reindeer House and we would scrub the floors and make everything tidy for her imminent arrival. And on the day of her departure we would have big smiles on our faces and with a cheery wave as she drove away we could look forward to having the house to ourselves again.

Family life with one and then two young children carried on alongside caring for Britain's only herd of reindeer. Alex and Fiona had no choice but to be involved from a very early age. Initially on my back in a baby carrier, then as soon as they could walk joining us on the hill while we handled and herded the reindeer, and they dodged reindeer who chased them. We were truly a reindeer family. And then Dr Lindgren passed away in her eighty-fourth year and with all the knowledge we had learnt over the previous ten years we were convinced we could make a go of earning our living from this unique herd of reindeer.

• 9 •

TAMING THE WILD

JOURNALISTS LOOKING FOR a story often ask me, 'What is it that you most like about your reindeer?' For me it is an easy question to answer: 'It's their wildness but also their tameness.' Wild because they live and thrive in such a hostile, unforgiving place but tame because they have been domesticated by man for thousands of years.

We handle our reindeer a lot, whether it's to administer medicines, wormers and vaccinations or to train them to be led on a halter or to be harnessed to a sleigh. We regularly lead reindeer down off the hill to Reindeer House and when we are moving the herd we will halter one reindeer as a leader. This handling from when they are very young makes them tamer than many the semi-domesticated reindeer of Scandinavia and Russia. And without a doubt our reindeer are even tamer for being free to roam and not held permanently captive. Indeed, the Sami reindeer herders have a wonderful saying: 'If you give the reindeer their freedom they come back to you.' And this is so true.

So how did the taming of reindeer come about? In Europe the history of reindeer and man is particularly rich because most ancient reindeer remains are associated with primitive human cultures. Modern humans arrived in Western Europe about 35,000 years ago when Upper Palaeolithic man displaced the more primitive Neanderthals and brought with him more sophisticated tools made from wood, stone and bone. Man's existence was intimately bound up with the animals and plants around him as he hunted, fished and collected shellfish, grubs, wild fruits and vegetables.

Upper Palaeolithic people seem to have depended heavily upon reindeer as a source of food and clothing. Kill sites containing many reindeer bones suggest that the animals were extremely important to them. Indeed, at a number of kill sites over 75 per cent of the total bones present were reindeer. Reindeer also feature heavily in early cave art and rock drawings. About 13,000 years ago, reindeer featured in Palaeolithic art as far south as Altamira in Spain. The seasonal movements of reindeer were also evident at this time, with the human cave dwellers of the Pyrenees following the migrating herds to the Atlantic and Mediterranean coasts.

All this activity was taking place at a time when glaciation had reached its maximum in Europe, and so the southerly distribution of reindeer was at its most extensive. This led French prehistorians to recognise this era as 'L'age du Renne', suggesting that this was a period of time

Reindeer antler is smooth on the surface and very dense. By hand or mechanical sanding, an already smooth surface can become incredibly polished and almost feel like ivory.

when reindeer were of the utmost importance to man. In many respects reindeer were the ideal prey. Their seasonal movements were predictable, using the same crossing points over rivers, calving grounds and paths every year, and their curiosity made them easy prey. Their meat provided food, their antlers could be used as tools and their skins could be made into waterproof and warm clothing and tent material.

The domestication of animals in general by man is a process generally associated with Neolithic times (5,000–2,500 years ago in Europe), when man gradually moved from hunter-gathering to farming. During this transition from a nomadic lifestyle to a sedentary one, a number of our common domesticated animals of today were at the beginning of the process of being enclosed by man and so removed from their natural environment. Crop-robbers like cattle and pigs were confined to the human environment and bred in captivity. Cats, which entered human encampments in pursuit of vermin, were selectively bred to produce the domestic cat of today, and in a secondary form of nomadism, after the early sedentary farmers had destroyed their immediate environment, horses and camels were domesticated so man could revert back to a nomadic existence, where they moved with the seasons to find food.

Even before this transition to settled life, there is definite evidence to suggest that some domestication of animals was taking place while man was still basically hunter-gathering. Wolves and man occupied a similar niche and had long pursued similar prey, and it is likely that at least 15,000 years ago the complex social systems of man and wolf teamed up in some loose association to improve hunting techniques to the benefit of both. From the wolf the dog evolved as a hunting companion to man about 6,000 years ago.

As with the wolf/dog, it is quite possible that long before the advent of agriculture there were loose associations between Upper Palaeolithic man and migrating herds of reindeer – associations that could perhaps be regarded

Recent research has shown that reindeer, unlike most other mammals including humans, can see ultraviolet with no harmful effects. This is thought to be a good survival strategy for reindeer. Urine, lichen and fur look black in UV light and so reindeer can see urine from other animals and by smelling this can detect the presence of other reindeer or indeed a predator. Lichen is an extremely important winter food and so again is easily detected in the scattered light. The fur of wolves, the main predator of reindeer, may look camouflaged with normal vision but to a reindeer with UV vision it stands out black.

as the roots of reindeer domestication. Reindeer are certainly predisposed to domestication, being highly gregarious animals that tolerate the presence of predators like wolf and man shadowing the herd on the move.

Various theories have been put forward to explain how the domestication of reindeer came about. Initially there must have been some form of social contact between man and beast, perhaps as a result of man hunting his prey. The social contact between man and reindeer could well have been accentuated during the winter months when reindeer are in a state of deprived nutrition, feeding mainly on lichens and eating snow for water. They are known to suffer cravings for minerals and salts during winter, and these cravings could well have been put to good use by man. It is well known that reindeer are attracted to human urine, and indeed this is thought to have attracted and bound them to human camps. Supplying salt for them would certainly have emphasised this bond and increased the social contact between man and beast.

Most forms of animal domestication, like cattle, sheep and dogs, resulted in the animal being removed from its natural habitat, enclosed and ultimately the fully domesticated animal bearing little resemblance to its wild ancestors. But in the case of reindeer, man and reindeer alike have remained in a nomadic state. This is essential to the well-being of the reindeer because of their highly specialised Arctic adaptations. In order to utilise the reindeer as a domesticated animal, man had to accept the animal's natural requirements and allow its nomadic lifestyle to continue.

The reindeer that were attracted to the human camps, even if few in number, could well have helped man to hunt wild reindeer more effectively. It is likely that the early stages of domestication came about as a result of decoy hunting. Various examples of decoy hunting of reindeer have been recorded over the last few hundred years. The ancient Samoyed reindeer herders, for example, hunted wild reindeer by selecting four or five young females from their herd; attaching them to ropes, they would approach a herd of wild reindeer under their cover until the hunter was close enough to shoot his arrow. Another ancient reindeer-herding people, the Tungus, adopted a different method. They would leave a few of their tame females on the feeding grounds of wild reindeer during the rutting season, and when the wild males were busy trying to mate with them the Tungus would creep up and shoot them. Another method was to release tame mature bulls during the rut with rope tied round their antlers. During the inevitable tussles with the wild bulls their antlers would become hopelessly entangled and the captured wild reindeer could then be killed.

Driving herds of reindeer into elaborate pitfalls, traps and stockades was an extremely productive hunting method. Pitfalls were dug on migratory routes at places where the trails were restricted, like narrow valleys, and there is evidence to suggest that movement was further restricted by extensive fences guiding the animals into lines of pitfall traps. Thousands of examples are known, the most extensive ones having as many as 500 pitfalls over a distance of 5 miles. Stockades were another very effective form of hunting on a large scale. Fences radiated out across open ground and were used to drive large herds of deer into a circular enclosure; once inside the corral they could be either killed or incorporated into already tamed herds.

The tame reindeer were of paramount importance to these processes, and would have been a valuable asset to the hunter. In return, he would protect his tame reindeer from danger and perhaps actively encourage the tame females to breed with wild males to produce young that would grow up tame. Unintentionally, the hunter would have become a reindeer breeder, but without making any profound changes to the physiology and appearance of the animal. The advantage to a tamed reindeer lay in the protection offered to them by man from predators.

The process of domestication of animals often led to the persecution and eventual extermination

of the wild ancestor. However, in the case of reindeer the process of domestication proceeded hand in hand with the continued hunting of wild reindeer. Indeed, the tame reindeer was a vital team member in the act of hunting his wild cousins.

Unlike the reindeer of Eurasia, the caribou of North America were never domesticated. They would undoubtedly have been of prime importance as a hunted animal in Palaeolithic times in North America, but the hunter-gatherer lifestyle appears never to have evolved further, and to this day the indigenous people of North America are hunters rather than herders.

Until the 1950s, when they moved to permanent settlements, the Chipewyan Indians ('caribou-eaters') maximised their contact with caribou by moving from winter settlements in the boreal forest to summer camps near the tree line – a lifestyle reminiscent of the

When reactor 4 of Chernobyl's nuclear power station exploded on 26 April 1986 the reindeer herds throughout Arctic Scandinavia were acutely affected. This is because the lichens that the reindeer were feeding on had accumulated the radioisotope caesium-137 up to 165 times higher than levels ever recorded. The consequences for the Sami were severe as hundreds of tonnes of reindeer carcasses had to be disposed of as toxic waste.

herd-following practices of the early domesticators of reindeer in Eurasia. It appears to have been a successful strategy for survival as the Chipewyan have few legends of starvation, in contrast to the sedentary Inuit on the Arctic coastline, who suffered terribly in times of low caribou numbers.

Caribou and wild reindeer today can almost be regarded as a different breed from the domesticated form. As with all truly wild animals, any attempt to tame individual caribou or wild reindeer produces the same problems of stress and agitation that would be found in any wild animal not used to human presence. This has led to a school of thought among Russian scientists that the ancestors of domesticated reindeer today perhaps came from a particular 'race' of reindeer in Eurasia that was predisposed to the process of domestication. Indeed, the whole process of domestication is thought by some to be an active process on the part of the animal. By associating themselves with man they passed on the responsibility for escaping their main predator, wolves, thus forming a partnership to the benefit of both sides.

However, the fact that the wild and domesticated forms are so different in their behaviour today could equally be explained by the fact that the process of domestication spans many thousands of years and is not completed in just a few generations. Perhaps the cultures of the Native Americans merely precluded any concept of nurturing the prey, protecting them from predators, or capturing them in corrals to tame them or harvest the meat.

· 10 ·

REINDEER:
THE ARCTIC'S
FARMED ANIMAL

HUNTING, HERDING AND following herds of reindeer provided the opportunity for man to utilise the reindeer as a source of meat and skins, but following on from this the reindeer in time became a beast of burden. Ridden on, carrying packs and pulling sledges – they became the farm animal of the north.

Paintings, thought to be 3,000 years old, on the walls of caves beside the River Lena in Russia show humans walking among reindeer but without weapons, suggesting that man was by this time beginning to manipulate and 'herd' the animals. Also thousands of rock paintings and engravings discovered in the Alta Fjord, north Norway, depict various scenes of human activity dating from 4,200 to 500 BC. As well as hunting and fishing scenes, large groups of reindeer are depicted with fences corralling them.

More concrete evidence of reindeer domestication comes from later written accounts in Chinese chronicles. Yao Silian, in his *Chronicles of the Liang Dynasty*, written between AD 629 and 636, wrote about people living in the northern forests who kept deer instead of cattle and used them to pull 'carriages'. In the *New Book of the Tang Dynasty*, written between 1044 and 1060, mention is made in more detail of a tribe called Ju who lived north of Lake Baikal, 'not having sheep nor horses, but deer', and using the deer to pull carriages.

(There was no word for sledge in the Chinese language at that time, and 'carriage' would have been the closest description.)

Certainly it is likely that more sophisticated stages of domestication came about when the reindeer herder came into contact with other herding people or with settled farming folk. It is unclear exactly where and when reindeer were first packed and ridden, although both the Tungus living east of Lake Baikal and the Sayan on the border between Siberia and Mongolia have been described as pioneers in the art of reindeer packing and riding. Whether the two regions developed the practice independently is open to dispute and various contradictory theories have been put forward, based on studies of techniques, styles and equipment. In the mountains of Sayan drawings have been found that indicate man was riding reindeer possibly 2,000 years ago.

From the same region comes indirect evidence for the riding of reindeer in the first millennium BC. The frozen remains of horses were found in the Altai Mountains of south Siberia, and one of the horses was equipped with an elaborate mask that appeared to transform the animal into a reindeer with antlers. This 'funeral horse', complete with saddle stuffed with reindeer hair, is thought to be a very superior sort of riding reindeer. The practice of riding reindeer was probably derived from horse culture, since the saddles used for riding reindeer in south Siberia are very similar to horse saddles. Pack saddles for these reindeer were almost definitely derived from the same origins.

However, the ancient Tungus have also been credited by some as the original domesticators of reindeer. The Tungus are the ancestors of many different groups of reindeer hunters and herders in Russia today, including the Evenk, Even and Dolgans. Their extensive spread from their original territories east of Lake Baikal is an indication of how successful they were. Like the Sayan, it is very likely that they developed a method of riding reindeer to enable them to travel with ease through the taiga, to allow them to hunt and fish. However, their style of riding was fundamentally different from that of the people of the Sayan: instead of riding on the backs of their reindeer they 'perched' on their shoulders. For the reindeer this was a much better technique, as they have a relatively weak back compared with horses. In addition, the Tungus mounted their reindeer from the right instead of the left side, and used a long pole for balance instead of stirrups. The Tungus' method of riding was definitely superior and allowed them to spread far into the Siberian taiga.

Packing reindeer for carrying loads almost definitely preceded the technique of riding them and for the Tungus and the Sayan the riding saddle was

most probably an adaptation of the packsaddle. However, packing reindeer were also widespread among the reindeer-herding people of northern Europe. The Sami, for example, never rode their reindeer and their type of packsaddle was completely different from those used in south Siberia. It is therefore generally agreed that the two types of reindeer packs evolved independently.

The use of reindeer for pulling sledges is almost universal among the reindeer tribes of the tundra and was almost certainly learnt from dog sledging. The harness used by reindeer herders in east Siberia has strong similarities with a dog harness. Riding reindeer would have been the most practical mode of transport in the boreal forest, but sledging would have been more compatible with the wide-open spaces of the tundra. Indeed, as draft animals reindeer were far superior to dogs not only because they could pull heavier loads but also because they could literally 'feed off the land', grazing from the natural vegetation along the way.

> Reindeer's perfect adaptation to a harsh arctic environment is of the greatest benefit to humans. Humans would never have been able to inhabit the Arctic without the reindeer but the reindeer can live in the Arctic without the humans. This makes humans a parasite of the reindeer.

As beasts of burden, the reindeer opened up vast areas of otherwise inhospitable land to man. The tundra and taiga could now be utilised by nomadic people traveling easily over long distances to hunt and fish.

Direct suckling of milk from lactating females by herdsmen and children was probably an ancient custom among early herding people but hand-milking and cheese-making were probably learnt much later. The reindeer people of north Scandinavia probably copied the practice from Scandinavian farmers because their words for milk, cheese and milking implements are of Germanic origin. Milking was also developed, probably independently, in northern Asia, where again it was probably copied from cattle farmers.

By associating themselves with an animal as superbly adapted to its environment as the reindeer, man was able to move into some of the most inhospitable areas of the world.

Domesticated reindeer were a vital part of the hunter-gatherer lifestyle that reindeer people pursued. They used the animals not only to access their hunting and fishing grounds but also to carry their belongings as they made their seasonal moves. Only a few reindeer were needed for these tasks but they were so important for this role that they were not regarded as a source of food. Only wild reindeer were killed for their meat and skins. But this modest lifestyle was gradually to change as there was a shift towards keeping larger groups of domesticated reindeer. This seems to have happened independently in many different areas of north Scandinavia and Siberia but a lack of archaeological evidence means the exact timings of the rise of reindeer pastoralism before 1600 is unclear. We do know that it took place some time before 892, because in that year the Norwegian chieftain Ottar sent a letter to King Alfred of Norway informing him that he owned 600 domestic reindeer.

Certainly by 1600 a system dominated by hunting and fishing but supplemented by small-scale reindeer breeding was prevalent among most native people of the Eurasian tundra. Domesticated reindeer were still exclusively used for transportation, with hunted wild reindeer providing meat and skins for food, household needs, tents and clothing.

It was not until the early eighteenth century that numbers of domestic reindeer began to grow, particularly among the tundra herds, and in the mid to late 1700s there was a population explosion, with herds increasing tenfold in some cases. Wealthy herders owning over 1,000 reindeer were compared with less wealthy ones having 100 to 200 and poor families with about 20 to 30. As the domestic herds of reindeer grew, so wild reindeer numbers fell and for many of these people the hunting of wild reindeer virtually disappeared.

There are various theories as to why the domestic herds suddenly blossomed over a period of about 150 years. Any increase in numbers would have depended on ecological as well as social factors. Even though the reindeer had been domesticated, they were still totally reliant on the climatic conditions and the quality of natural pastures. Hot dry summers in the tundra are particularly difficult for reindeer, since the resulting poor vegetation means they face the harsh winter in poor condition. And although reindeer are well adapted for temperatures as low as -25 to $-30\,^{\circ}$C, winters severer than this and with drastic temperature fluctuations take their toll on conditions and raise natural mortality. Since wild and domesticated reindeer are physiologically very similar, it is evident that the two populations would thrive or fail under the same natural conditions, and so in this instance the two populations would have increased simultaneously. The knock-on effect of this would be greater numbers of wild reindeer available for hunting, which would alleviate the need to slaughter any domesticated animals. Indeed, many herders, despite owning many reindeer in some cases, strenuously avoided slaughtering their own animals even when they were faced with starvation conditions from lack of animals to hunt or fish.

Some herders regarded the hunted or fished animal as their food and their owned reindeer as their wealth. Only in an emergency would they kill a domestic reindeer. And when they were forced to slaughter a domesticated reindeer, it had to be done in such a manner as not to spill any blood, otherwise it was considered sinful! The meat from game animals and particularly wild reindeer was considered to be more prestigious than that of the domesticated reindeer, to which many herders were indifferent. The tradition of conserving as many domestic reindeer as possible was deeply embedded and probably helped to accumulate huge herds of domesticated animals, which would eventually evolve into a food-producing reindeer economy.

The transition to a meat-producing economy took place simultaneously among many different tundra nomads, from the Sami in the west to the Koryak and Chukchi in the Far East, where both groups amassed large herds. Although this accumulation of domesticated reindeer was partly a response to demands for meat, it could also have been because of a need for reindeer skins. Meat could be obtained from hunting game but there was virtually no substitute for reindeer skins. And not just any old skin! Skins from reindeer slaughtered at different times of year and at different ages were all required to meet the demands of the herders' everyday life. A special slaughter in July produced the short-haired skins needed for the inside set of winter clothing, while early winter kills provided the thicker coats needed to make inner sleeping chambers for tents. Skins from the latest slaughter in October were necessary for the very warmest clothes, needed when winter hunting for sea mammals – a practice that involved long hours sitting out on the ice. The lack of any of these particular types of skin would have jeopardised the very existence of these peoples.

Whatever the reason for the increasing size of domestic herds, harvesting meat from a herd of domesticated reindeer was a much more reliable source of food than hunting wild reindeer and led to increasing numbers of indigenous people. In turn, this led to increased competition for grazing with the wild reindeer. They still led a nomadic life, but now the herders followed a specific herd of reindeer as opposed to hunting only the wild ones.

Any surplus of reindeer meat and skins meant the herders could trade and barter with other people. In the early 1800s the Komi reindeer breeders were known to kill regularly large numbers of reindeer and use the meat and skins to 'buy' goods such as flour or butter and to pay hired labour. The Chukchi and Koryak also began to trade with the coastal communities, supplying them with reindeer meat and skins. The coastal hunters suffered from the declining numbers of wild reindeer,

Chukchi legend: A long time ago Kutkh the Raven was flying around on his own business. In his claws he held an enormous rock. Finally he could hold the rock no longer. He let go, and the rock fell to earth, splitting into many pieces. The biggest chunks formed the mountains and hill, the medium-sized pieces the reindeer. From the small ones came the trees and shrubs, and from the dust came the people.

which all but vanished from the far east of Russia as the domestic herds grew. The size of the domestic herds varied, people of the tundra tending to keep larger numbers of animals than the people of the boreal forest. No doubt this reflected the more solitary habits of the forest reindeer compared with the more gregarious tundra reindeer. Regardless of the ways in which these different people used their reindeer, with them man could live and thrive in the Arctic and sub-Arctic areas of the world. The reindeer became their cows, sheep and horses – three animals for the price of one!

· 11 ·

A HERD OF INDIVIDUALS

FROM THE VERY beginning, all of our reindeer have been given names. This was a practice
initiated by Mr Utsi and in the early days many of
the reindeer were named after people and places
Utsi and Lindgren knew. Sarek, the trusted ox who
came in with the first consignment, was named
after one of the highest mountains in Swedish
Lapland. Aviemore and Murjek (the winter home
for Mr Utsi's herd in Sweden) were breeding bulls
who came at the same time as Sarek. But as the
herd grew that became more tricky (maybe they
had run out of friends who wanted a reindeer
named after them) and they struck on the idea of
naming calves born each year after a theme.

Since 1971, a different theme has been chosen
including different species of trees, whisky and
wines, characters from books and even well-known
authors. In 2002 it was 50 years of reindeer in
Scotland so we celebrated with gold: Ingot, Finch,
Bangle, Sequin. Then in 2012 with the London
Olympics, the Queen's 60th Jubilee and our own
celebration of 60 years of reindeer, we chose names
like Olympic, Mo, Duke, Balmoral, Second and
LX, all real characters in the herd today.

To the untrained eye a group of reindeer may all
look rather similar but to the reindeer herder who
sees them frequently the different tones of colour,
shape of body, antlers and head and their individual characters all help to distinguish and name

them individually. We also ear tag all the reindeer with a different colour for each year, which certainly helps the rookie reindeer herders!

Sometimes, though, a reindeer is 'different' enough to warrant a name outwith the theme and two especially spring to mind. Mystery was quite literally a mystery! Born in May 1990, as a young calf with his mother he would have headed into the Cairngorm range for the summer. Keeping regular checks on the free-ranging cows and calves is never easy and inevitably what goes out for the summer doesn't necessarily come back in the autumn. Then one day in early autumn Alan found a note on the windscreen of his reindeer van saying there was a 'baby reindeer' wandering about in the ski carpark. Sure enough, when Alan got there, there was a calf, a few months old, standing alone in the middle of the carpark, looking very lost. With the help of a couple of walkers who had just come off the hill, Alan managed to catch the calf, bundle him into the van and drive down to Reindeer House. The following morning we presented the wee chap to a female reindeer, Tiree, who had arrived back from the free range without her calf. But she just gave us a haughty look as if to say 'that's not my calf'. So we named him Mystery and it wasn't until the end of the summer when all the reindeer cows had returned with their calves that we managed to work out by a process of elimination that his mother

must have been our oldest cow Natasha. She must have succumbed over the summer, leaving Mystery to find his own way home. Our conclusion was confirmed when Mystery grew up to look exactly like his mum.

One of the trickiest things to do is hand-rear a young, motherless calf. Apart from the obvious problems of finding a substitute for the rich reindeer milk, the orphaned calf tends to become a bit of a misfit, often being shunned by the rest of the herd, and generally regarding itself as partly human. Jura, Beauty, Dubh and Utsi were all successfully hand-reared reindeer. They probably survived because their mothers were able to raise them for at least the first week. Reindeer colostrum and the first few days of rich milk provide a great boost for any reindeer calf.

So it was a bold step for us to take a newborn calf away from his mother when he was just hours old. Marie, his mother, was notorious for losing her calves when they were just a few days old. We suspected she didn't produce enough milk. This had happened once too often, so we decided to 'bite the bullet' and try to raise her next calf ourselves. Taking her calf away was traumatic for Marie. Reindeer are good mothers, and for days afterwards she would challenge us on the hill, as if to say, 'What have you done with my calf?' It was up to us to prove to her that we had done the right thing. We rather perversely decided to call the calf Fred, after the rather comical film in 1991, *Drop Dead Fred*.

Goat's milk is the best substitute for reindeer milk, and every two hours, night and day, Fred drank his milk. He never looked back, progressing to fewer bottles of milk supplemented with lichens and leaves. He lived in an outside pen beside the back door (reindeer calves will overheat if you bring them inside). Our daily routine all that summer was regulated by Fred's feeding times. By the time he was a month old we thought it was time for us to have a break, so we left him for the afternoon and went off to the local Highland Games at Newtonmore.

Of all the deer species, reindeer has the highest amount of solids (protein, fats, mineral and carbohydrate) in its milk to ensure rapid development of the calf.

Unfortunately, Fred's first afternoon alone could not have been worse. While we were away there was an almighty thunderstorm with torrential rain. When we returned from our afternoon out Fred was drenched and exhausted; he had only enough energy to stumble through the door into the house and collapse on the floor in a heap. That night we resigned ourselves to the fact that Fred was going to die. Reindeer are not great fighters, and when the chips are down they are quite happy to turn up their toes and give in. As the evening wore on there was very little change. However, at about 10 p.m. Fred uttered a grunt, indicating that he was hungry. We couldn't believe it: perhaps Fred was going to survive. With some warm milk inside him, life seemed to return to his limp, cold body. It seemed that Fred wasn't going to give up that easily.

Happily, Fred recovered completely from his drenching but soon he had to face another threat to his life. It transpired he was born with a hernia. If the vet didn't operate on it while he was still young, it could well lead to serious complications when he was older. Having invested a great deal of time and emotion into this young reindeer, it seemed only sensible to give him the best chance for the future, so we agreed to the operation. General anaesthetic for reindeer is not exactly a common request, and on the few occasions it has been used there was great concern as to the amount needed. Too much, and they'll never wake up. The operation was performed on the kitchen table at Reindeer House, because 2001 was the year of the foot-and-mouth outbreak and there was a ban on moving cloven-hoofed animals in a bid to stop the disease spreading. Leaving our two vets, Andrew and Jane, in charge, I decided to take a walk. I couldn't bear to watch, and there was nothing I could do after we had manhandled an unconscious Fred up onto the makeshift operating table. When I returned the job was done, and Fred lay flat out on the floor in the sitting room. Now we just had to wait for him to wake up.

A few hours went by. We had to step over Fred's prostrate body on the floor, while the dogs circumnavigated him. He continued to breathe and his eyelids twitched, but nothing more than that. Eventually, to try to jolt him back to consciousness, we decided to offer him a bottle of his favourite goat's milk. To our delight his nose twitched and with some difficulty he raised his head to reach for the teat. Hunger had got the better of him, and it would just be a question of time now before he was back on his feet again. Despite a couple of close calls, Drop Dead Fred had survived to tell his tale. Perhaps there is something in a name!

In fact, it was 'Fred the Baby Reindeer' who graced the front cover and middle-page spread of the *Daily Telegraph Magazine* during Christmas 2001. This followed a visit from a journalist who I took on the hill to see the reindeer. As the reindeer approached, looking for their food, I was able to introduce her to some of the characters, naming them as I did. Some of the reindeer stood patiently waiting to be fed, while the odd greedy ones nuzzled the bag looking for a way in. As we led them away to a suitable place to put the food

down the reindeer walked alongside us, like we were part of the herd. Once back at the house for a cup of tea, Fred – who was still a calf – made his entrance into the kitchen, looking for his bottle of milk. The lady was gobsmacked by everything she had seen. Later, when I read the article, she described the outing as if she was at a cocktail party and I was introducing her personally to the guests. It was the perfect analogy.

Not a day goes by when an individual reindeer is not mentioned for some reason, maybe for their good or bad behaviour, loss of antlers or state of health. The herd is truly a group of individuals but within the herd there is also 'The Sisterhood' – small family groups of related females: mothers, daughters and sisters who can often be seen sticking together, maybe even looking out for each other.

Back in the early 1990s I decided to trace the current reindeer in the herd back to the original reindeer that arrived in Scotland. This was possible because of the detailed breeding records that Dr Lindgren insisted we keep. I only attempted to do this through the mother's records, as nobody can ever be absolutely sure who the father is. Indeed even now, when we try to ensure certain bulls rut with particular groups of females, best laid plans can be undermined by the unexpected. The rut of autumn 2011 was, we thought, well planned: five young bulls, Springbok, Macaroon, Magnus, Bannock and Caribou, would be the main breeders, chosen for their new blood, colour markings and character. Each one ran with half a dozen cows and the cows we didn't want to breed from were put out to the free range, away from the breeding bulls for the duration of the rut.

All the young bulls not used for breeding were separated as a single group. One of these young bulls, Strudel, however went missing and despite searching again and again over several days, he was nowhere to be seen. With a lot of rough ground, hidden depressions and in places dense woodland we assumed he must have died for some reason. You can imagine our relief when three weeks later he showed up on the free range. But to our dismay he was with all the cows we didn't want to have calves next year! Strudel had found his own 'heaven' and the following spring we had a bumper calving with lots of little Strudel lookalikes!

Since the original reindeer came to Scotland there have been many generations of reindeer and looking back over the family trees it is interesting to see how productive, or unproductive, different females have been. Vilda, a young female brought in to Scotland by Mr Utsi in his third consignment of reindeer in 1954, was the 'mother of all reindeer'. Over her breeding life she produced eleven calves, and nine generations later we still have reindeer born who are directly related to her. That family produced a series of big females, some more productive than others. Part of Vilda's family tree is featured at the end of this book.

Another early import was a female reindeer called Assa, who was a dark-coloured female, but her only offspring Willow was very pale and today many of the 'white reindeer' in the herd are related to her. Snowflake was Willow's granddaughter and she was actually pure white, which we discovered many years later is linked to deafness, and the pure white reindeer we have in the herd today, Blondie and Blue, are both stone deaf. White reindeer are highly prized in Sami culture. Traditionally, white reindeer were regarded as lucky and Mr Utsi always longed for more white reindeer in the herd. Indeed he had a rather wacky idea which involved painting rocks white to encourage more white calves to be born! But not so lucky for white reindeer as their skins are highly prized when traditional outer garments are made out of reindeer skin. White reindeer are also often described by the Sami as 'lazy' and easily killed by predators. What amazes me is that until I told a Sami one day, he didn't realise these pure white reindeer are actually deaf! Something else weird about white reindeer is that their hooves grow quicker than normal-coloured reindeer. When reindeer are free-roaming and covering many miles of rough ground in any one day, their hooves naturally wear down but the white reindeer do appear to have rather long toes in comparison. Interestingly, I had a discussion along these lines with an alpaca farmer and he confirmed that white alpacas also grow their hooves faster.

Two strong family traits are tameness and greediness. Within family groups you tend to find tame, greedy mothers produce calves with similar traits. Hopscotch is a great example: an incredibly greedy cow (I mean that in the nicest possible way!), both her calves Kips born 2015 and Kipling two years later have their mother's same voracious appetite, which they would have both inherited and learnt. If a calf watches his or her mother always beside the feedbag looking for more food then they could very quickly learn to do the same. At the opposite end of the spectrum, wild mothers tend to produce wild calves, and for us the classic of all time is Lilac who has to be one of the wildest and most suspicious females we have ever bred. Two of her female calves, Wham and Weevil, definitely learnt to be wary like their mum and would always cause mayhem when we were herding a group of reindeer, dashing off in every direction rather than the one you wanted them to go in! There is always a sigh of relief when it comes to moving the herd and troublemakers like that are not around.

· 12 ·

ASSOCIATION OF WORLD REINDEER HERDERS

ALTHOUGH OUR HERD is of modest proportions and our reindeer culture can be traced back to less than a few decades, in September 1993 we were made to feel we really belonged, if only in a very small way, to the wider world of reindeer people.

Along with nomadic people from right across the northern hemisphere, we were invited to the First International Gathering of Reindeer Herding People. Subsidised by the Norwegian Government and affiliated to the Fifth World Wilderness Congress, which took place simultaneously, the gathering met in the Arctic town of Tromso and was attended by representatives of the majority of indigenous reindeer herders.

By no stretch of the imagination could the Smiths of Reindeer House be described as a 'reindeer tribe' but we do nevertheless live by our reindeer and we had no hesitation in accepting the invitation. Armed with cameras, warm clothes and our quota of malt whisky, Alan (who had never flown before), my father and I took off from fogbound Aberdeen to begin the trip of a lifetime.

The west coast of Norway looked truly spectacular as we flew north from Bergen to Tromso over sea, fjords, mountains and forests. Tromso is a most attractive place, situated on an island and surrounded by massive fjords, disappearing into mountains that rise above 6,000ft. The tops of the mountains had permanent snowfields and looked magnificent. The autumn air was crisp and clear, with a tremendous Arctic freshness and being within the Arctic Circle there was constant daylight in the summer months.

The venue for the opening ceremony of the Reindeer People's Gathering was the local Fokus Cinema and representatives from all the reindeer breeding areas were lined up on the stage for a photo call. I was the representative for the Scottish contingent.

The colourful costumes of the people up on stage looked wonderful. The Sami people's dress, different styles representing different districts, included heavily embroidered tunics and ornate headdresses. Many of the Russian reindeer people, Chukchi, Evenks, Nenets and Evenkis were predominantly in costumes made from reindeer skins, incorporating the contrasting light and dark colour types. The borders of the coats were decorated with brightly coloured red and yellow felt. With long coats, skin boots and brightly coloured headdresses wrapped firmly round their faces, they looked ready to face even the longest winter of snow and sub-zero temperatures. I did feel a bit of a fraud standing next to these people who represented many thousands of years of reindeer culture. Their knowledge came from tradition passed down through generations; mine had been learnt in a brief few years.

Representatives from each region gave a short presentation about their way of life. The picture they conjured up was one of a vast expanse of mountains, tundra and taiga pastures, with relatively few people herding huge numbers of reindeer. The land extends unbroken from the west coast of Norway, above and below the Arctic Circle and halfway round the world to the Bering Strait. Over the entire area there are probably 3 million semi-domesticated reindeer being herded and slaughtered for their meat, skins and antlers. A far cry from the Cairngorms.

The festival included a lassoing competition, not with the live reindeer but antlered skulls screwed onto pieces of wood. Lassoing among reindeer herding people is taken very seriously and is an important tool when catching individual reindeer in a corral of moving animals. It is very precise and with practice the thrower can quickly pick out the reindeer they want to catch. A traditional lasso is crafted from leather and sinew, with antler as a sliding

200-year-old Russian reindeer leather was found by scuba divers when they discovered the wreckage of the *Catharina von Flensburg* in the Plymouth Sound. The cargo in the muddy seabed was remarkably well preserved bundles of reindeer leather hides, which had been used to make bespoke and very expensive leather goods.

toggle. Many of the Russian herders used traditional ones, but the Sami had lassos made of plastic-coated rope, with a plastic toggle.

At night there were concerts of traditional Sami joiking (singing with no music accompaniment) and the banquets were a carnivore's paradise, each time revolving around lumps of reindeer meat: boiled ribs, tongues, marrow, even eyeballs, usually served in a thin meaty (and salty) broth. The accompanying vegetable was always potatoes and the dessert cloudberries with cream. Cloudberries look like apricot-coloured raspberries; they are found all through the Arctic on boggy ground. We do have some in the mountainous regions of Scotland but not as abundant as in the Arctic.

After spending three days in Tromso and seeing only one reindeer, who had been brought into town by a Norwegian Sami, we were keen to get out into the wilds and find out what real reindeer country was like. A bus took us from Tromso to Kautokeino, over land where reindeer normally spend the winter and so the only herd of reindeer we saw was when we were taken to a slaughter house. A herd of about 200 reindeer were milling around in a back corral and from there they took them through the killing unit. It wasn't quite what we expected but equally incredibly interesting to see how quickly an animal was despatched, skinned, gutted and finally the carcass hung alongside dozens of others. Reindeer meat is eaten fresh, smoked, dried and, in the case of some of the Russian reindeer herders, raw – something we also experienced first hand when we were taken to a reindeer corral out in the middle of nowhere. By the time we got there the advance party of Russians had already caught and killed a reindeer and were in the process of eating the choicest pieces of meat still warm and for all we knew still pulsating!

Our bus journey ended in the Norwegian town of Kirkenes on the Russian border and despite scouring the countryside at every stop with our binoculars there were no reindeer to be seen. But through translators we had a great opportunity to talk to amazing people from across the Arctic, about how they lived by their reindeer and indeed how we made our living from our Scottish herd. We must have made an impression because at the closing dinner in Kirkenes Vladimir Etylen from Chukotka, the Arctic region in Russia adjacent to the Bering Straits, stood up and announced he wanted to make one final presentation 'to the people with least culture but most commitment' and he asked one of us to step forward.

I felt incredibly honoured and not a little emotional as I walked up, watched by all those knowledgeable faces, and was given a beautiful ivory talisman made of walrus tusk. Vladimir said he hoped it would bring good luck and lasting

prosperity to Scotland's only herd of reindeer. Twenty-five years on and our talisman, a small rotund man, a little like a Buddha is kept in a safe place. But every so often I pick him up and remind myself of our memorable first encounters with reindeer herders from across the top of the world.

From this gathering in 1993, the World Reindeer Herders Association was finally formed four years later when the reindeer-herding community met, this time in Nadym, Russia. It was here that it was agreed to meet every four years at a different venue, always in the heart of reindeer-herding territory. The Association is organised and represented only by people who are actively reindeer herding, not politically motivated or represented by government officials for the region where they live. There are many problems that face reindeer herders today, some of them similar to other pastoral herders, like yak herders and nomadic camel herders. Often looked at by the modern world as an outdated way to farm animals, the pastoral way of life is also often practised in areas of high nature conservation. As well as discussing the many problems facing these indigenous people, from predation of their reindeer to loss of grazing due to industrialisation, the congress is seen as a way to celebrate the reindeer-herding culture and showcase their way of life to the rest of the world.

The first few congresses were in far-flung places like Inari in Finland, Yakutz back in Russia and Kautokeino, North Norway, but in 2013 the invitation came announcing the 5th Congress was to be held in Genhe, Inner Mongolia, the very part of the world Dr Lindgren travelled as a postgraduate student to study the Reindeer Tungus over ninety years before. This is where our reindeer story began, where Dr Lindgren cut her teeth on reindeer, and so we were hell bent on attending this time.

We arrived in the city of Hailar, the very settlement where Ethel Lindgren first heard about the 'far away tribe'. Then we travelled by bus along the River Gen and up into the forested slopes of the Kinghan hills to Genhe. Land that was wild and remote when Lindgren rode on horseback was now open grasslands full of livestock and farm crops, and the forests were harvested for their timber. Genhe, which would have been no more than a remote outpost when Lindgren passed through, was now a bustling city of high-rise flats and 150,000 people.

We came to the congress prepared with the story of Ethel Lindgren and her expedition to the Tungus, nowadays called Evenki. The photos we brought with us caused a huge amount of interest among the Evenki there and a baby carried in one of the group photos turned out to be a very old lady attending the congress. There were ripples of interest from many of the Chinese because for them these were photos they had never seen before of their relatives. Because of the change in political climate after Lindgren left and the clamp-down on foreigners to China following communism, her field notes and photos are an extremely valuable record of life in the 1920s and early '30s – a way of life that clings on to this day.

The Evenki are forest reindeer herders, using their animals as pack animals and milking the cows. Traditionally hunters, this is a practice they can no longer undertake because of the strict laws of gun ownership in China. But the authorities do see these people as a minority who need support and in 2003 the village of Aoluguya was built for the reindeer Evenki to live in. What the authorities didn't appreciate though was the reindeer belonging to the Evenki still needed to roam to thrive, and following a few disastrous years trying to enclose the reindeer they admitted the Evenki and their reindeer needed to return to their traditional ways, nomadising in the forests to allow their reindeer to find

When reindeer grow their new velvet antlers they frequently tap the growing tips with their back legs. Careful not to damage the soft velvet skin, the reindeer tap the antler with the ankle joint, above the sharp hooves. Why do they do it? Perhaps they can feel the antler growing and so tapping helps to relieve the growing sensation.

the right food depending on the season. The village of Aoluguya still stands. I'm not sure if the inhabitants are reindeer herders any more but there is an extremely interesting museum dedicated to the Evenki. The city of Genhe is known in China as Reindeer City and in recent years it has become a popular destination for Chinese tourists. A number of the souvenir shops there have deer-related gifts, including gift wrapped slices of velvet antler, an aphrodisiac used in traditional Chinese medicine.

The Association is represented by twenty-eight different regions from Alaska, Canada, Greenland, Scandinavia, Russia, Mongolia and China. During the congress in Genhe it was proposed that Scotland should be represented too since the Cairngorm reindeer herd introduced by Mr Utsi in 1952 was a free-ranging herd, living in its natural environment. This was unanimously agreed and I think this has been one of the proudest moments of my life when I stood up in front of all the members to accept Scotland as the twenty-ninth member of the Association.

Which fits nicely with a comment made to our son Alex when he was in Swedish Lapland in 2011. Swedish Sami Mikel Utsi (a nephew of the late Mr Utsi) said to our son Alex one day, 'You must be Scotland's first indigenous reindeer herder, since you were born into the Scottish reindeer way of life.' Well I suppose all 'tribes' have to start somewhere!

· 13 ·

INCREDIBLE
JOURNEYS

R ADIO COLLARS ATTACHED to female Alaskan caribou have shown that they are capable of traveling 5,000km in one year, making them the most traveled land animal in the world. The urge to migrate is huge for these highly gregarious herds and after a winter surviving on very little in the boreal forests, the pressure is on to move north to the tundra and the Arctic coast for the very first emerging spring vegetation, and very importantly before they calve. Obviously a young calf would slow down the migration, so its imperative they get there while still pregnant. Once born, with the newborn still unsteady on its feet, the herds of females will remain in the calving grounds. But as soon as the calves have gained enough strength the herd will be on the move, constantly looking for fresh vegetation and making the most of spring in the Arctic. The migration south back into the boreal forest will then commence in the autumn, the journey hampered and made less frantic by the onset of the rutting season. One of the best-known herds of Alaskan caribou is the Western Arctic herd, a migrating herd with their calving grounds in the far north-west of Alaska. The Porcupine herd is another migrating herd of Alaskan caribou and their summer grazing borders the Arctic Ocean on the north coast of Alaska.

Indigenous reindeer herders, the Nenets form the largest indigenous group of the Russian North and are one of the world's great reindeer herding peoples who have come to personify large-scale tundra reindeer husbandry. Nenet herders and their families migrate over long distances, up to 1,000km annually between the summer and winter pastures. Following the natural instincts of their reindeer through the seasons, these people are almost constantly on the move, using their castrated reindeer to pull sleds that carry people, tents and belongings, with the rest of the herd, which may be many thousands, herded along as well to fresh pastures.

Reindeer, because of their tractable nature, were obvious candidates for introduction into various Arctic and sub-Arctic areas, both mainland and islands. Some were even taken to the southern hemisphere, to the islands of South Georgia, Iles Kerguelen and Isla Navarino in sub-Antarctica. One of the earliest mentions of reindeer being sent overseas occurs in a book written in 1555 by Olaus Magnus. In *A Description of Northern Peoples* he describes the natural history of reindeer and their usefulness to man. He also mentions them being sent to foreign countries, especially overseas, but never surviving long. In particular, he refers to six reindeer, accompanied by a married couple from Lapland, being sent to the Duke of Holstein. Sadly neither reindeer nor humans survived.

Reindeer were introduced to more than thirty different locations during the twentieth century, particularly on remote Arctic islands or areas restricted by glaciers. Here they were often left uncontrolled and were allowed to breed without restriction in the absence of humans or other predators.

When a small population of reindeer arrived on high-quality pristine summer grazing it was not uncommon for the herd numbers to increase exponentially in the early years, partly as a result of increased fertility among the female reindeer, in particular females who reached sexual maturity at just 6 months old and so had their first calf at a year old. One of the best-documented cases concerns St Matthew Island in the Bering Straits, where twenty-nine reindeer were introduced in 1944. Astonishingly, this group multiplied rapidly and had grown to more than half a million in just twenty years. This rise was followed by a dramatic crash in numbers to almost zero in the next six years. Although the crash coincided with a number of severe winters, the harsh climate was not thought to be the only factor, since reindeer on adjacent islands did not suffer the same losses. As with other grazing populations, the effects of overgrazing can be dramatic, and this can be particularly accentuated with reindeer, which require very high-quality summer grazing to prepare for the ravages of winter. Once the population of St Matthew Island began to destroy the summer grazing, the reindeer could not properly prepare for the winter and so considerable numbers died. The number of deaths was further increased by the destruction of slow-growing lichens that were integral to their survival over the winter months.

Different islands provided different ecological conditions for the introduced reindeer populations, and following an initial irruption in numbers some of the islands went on to reach an equilibrium between the numbers of reindeer and the carrying capacity of the grazing range, and to this day there are still reindeer on some of these islands.

One of the best-researched island populations of introduced reindeer was to be found on South Georgia, an island off mainland Antarctica. Brought in by Norwegian whalers on three separate occasions between 1911 and 1925, the reindeer provided a welcome respite from whale meat but also fulfilled a role as a sporting amenity. Their introduction followed in the wake of other animals introduced to serve the whaling community. Cows, sheep, goats and pigs were all brought in to supply meat and milk but they remained domesticated and never became feral. The rationale behind the reindeer introduction was that this was an animal that would require no time-consuming husbandry and management.

Some of the reindeer cows arrived heavy in calf, but of course the reversal of the seasons in the southern hemisphere meant that the calves were born at the beginning of winter and so perished. It took the reindeer just two years to adapt to the change in seasons, with the associated reversal of rutting, calving, antler cycle and moulting. There were early losses due to avalanches of snow, but the reindeer appeared to thrive. Left to their own devices and basically uncontrolled, despite some culling for sport and meat, the reindeer quickly consumed all the available lichen and the tastiest summer plants. They then resorted to eating the dominant vegetation, *Poa flabellate*, a species of tussock grass available in both summer and winter.

In 2010, following a long consultation regarding non-native animals and plants on South Georgia, the decision was made to eradicate reindeer, rats and anything else that had been introduced, whether on purpose or accidentally. Since 2013, in excess of 6,500 reindeer have been systematically removed by culling and live catching, with some reindeer being introduced to the Falkland Islands with mixed success. Rats were targeted at the same time using poisoned bait and they too are no longer found in this sub-Antarctic region. To date, non-native flora has also been removed using herbicides but the dandelion has evaded complete eradication and clings on.

The spread of Christianity across the world and the influence of missionaries on the daily lives and beliefs of indigenous peoples are well documented. Indeed, not even remote Inuit populations living on the Arctic coastline of Alaska escaped religious interference to try to better their traditional lives.

In the late nineteenth century the Inuit people of western Alaska were subsistence hunters who relied on a variety of prey to survive. They lived a relatively sedentary life on the Arctic coastline for thousands of years, sustained by seals, whales, sea fish and the annually migrating caribou populations. Changes in the pattern of caribou migration meant that fewer caribou came by the Inuit settlements and this, combined with a shortage of whale meat after whaling ships from Europe and America arrived, resulted in starvation conditions for many of these indigenous people.

Desperate times need desperate measures and so a plan was put to the American Congress by the Revd Sheldon Jackson. He proposed that the Inuit should give up their unstable and unpredictable hunting way of life and instead

Caribou and reindeer have the greatest circumpolar distribution of any living ungulate – 14 degrees west to 5 degrees east and latitudinally 46 to 80 degrees. As a rule they are generally bigger in the south than the north.

become reindeer herders like the Chukchi, their neighbours across the Bering Straits in Russia, regardless of the fact that this was totally against the Inuit culture and traditional way of life.

Following reconnaissance trips to Siberia and early experiments with a small group of sixteen domesticated reindeer that were shipped across the Bering Straits to the Unalaska Islands, 171 reindeer were brought over from Siberia to mainland Alaska during the summer of 1892. With the reindeer came four Siberians who would teach the Inuit how to care for the animals. An early stumbling block was the fact that the Siberian reindeer herders and Inuit hunters hated each other and fairly early on it became apparent that there would be little cooperation between the two. The Siberians were soon replaced by Sami reindeer herders, who had been promised a better way of life in Alaska.

Sheldon Jackson hoped that reindeer herding would become the mainstay of the Inuit economy, and certainly the value of the reindeer was reinforced by the discovery of gold in the area. Not only would the reindeer provide meat for the Inuit, but they could also be put to work as beasts of burden hauling freight through the forests to the gold mining communities.

The demand for draught reindeer for the gold mining was immense. Following orders from

Congress, in 1897 a boatload of 538 castrated reindeer, already trained to pull sledges, was shipped across the Atlantic from Norway to New York, for onward transport by rail to Seattle and then by boat to Alaska. With the reindeer came 118 Sami herders and their families to teach the Alaskan native people the ways of reindeer herding. Copious supplies of lichen were brought along to keep the reindeer on their accustomed diet for the duration of the voyage across the Atlantic. The Lapps cared for the reindeer diligently, and only one reindeer died during the crossing and that was due to injuries sustained in a fight. Once disembarked on the American mainland, the reindeer were transported by train to Seattle, but here disaster struck. The lichen supply had all but run out and in a desperate attempt to save supplies for the onward journey to Alaska, they took the reindeer to graze in the local city park. This had dire consequences. Reindeer do not adapt well to sudden changes of diet and that night several of the reindeer died. Things then went from bad to worse, and by the time the reindeer reached their final destination in Alaska there were only 144 survivors.

Sheldon Jackson's intention was that the reindeer should ultimately become the property of the native people, providing them with income and food. Over the first fifteen years the reindeer numbers increased sevenfold to more than 10,000, some 2,700 of which were owned by at least sixty

Inuit. The project was set up in such a way that the Inuit apprentices 'earned' their reindeer. During training, Inuit who stayed the course were loaned two female reindeer each year. The offspring of those females became their own personal property. More reindeer stations were built and the reindeer population continued to increase. The cultural impact of young Inuit hunters turning herders was immense, and as long as the reindeer could be kept close to settlements all year round there was little effect on the Inuits sedentary lifestyle. However, as the numbers of reindeer grew, the grazing areas close to the stations became overgrazed and the Inuit had to adopt nomadic ways of reindeer herding, moving with the herds to find winter lichen pastures and taking them to fresh summer grazing on the coast. This led to problems, as the Inuit were reluctant to pursue what was clearly a very different way of life on the move.

Despite Jackson's noble intentions, commercial exploitation and non-native ownership of the herds increased. In particular, the Lappish families that had stayed in Alaska were treated far better than the Inuit when it came to reindeer ownership. This preferential treatment and their good reindeer husbandry led to a number of Lapp reindeer herders accumulating relatively large herds of their own. By 1920 over half of the 300,000 domesticated reindeer were owned by non-natives.

Notorious among the private non-native owners were the Lomen brothers, who arrived in Nome, Alaska, in 1900 to make their fortunes from gold mining. Their interest in reindeer was sparked when 1,200 reindeer owned by Lapp Alfred Nilima were offered for sale. Lomen and Company bought the herd and forged ahead with expanding herd numbers and proposals to export reindeer meat and associated products. Between 1928 and 1930 the Lomens shipped some 30,000 reindeer carcasses along with thousands of reindeer skins for making gloves. It wasn't until 1933 that the protests of the natives were finally heard and following a confidential report it was recommended that the Lomen Corporation should cease trading. Congress provided funds to implement the Reindeer Act, which allowed for the compulsory purchase of all non-native herds and facilities and their redistribution to native owners.

During the height of the 'Lomen era' the Canadian government commissioned a survey with a view to finding a suitable site to establish reindeer herding for the Inuit in Arctic Canada. Two Danish botanists were hired and their final report recommended the Mackenzie River delta as the most suitable site.

Once again it was the Lomen brothers who benefitted most from the subsequent deal, which entailed the sale of 3,000 reindeer. These would be herded overland from Alaska, covering some 1,600 miles of uncharted terrain. The Lomen brothers hired the expert Andy Bahr, a Laplander who had come over to America with the original boatload of reindeer destined to work in the gold mines. It was estimated that the trek would take two years, and the route would go inland across the Brook range of the Rocky Mountains to avoid the herds of caribou on the Arctic Coast.

So it was that on Boxing Day 1929, the 3,000 reindeer, accompanied by loaded dog sleds, reindeer sleds, Laplanders and Inuit, set out from Kotzebue Sound on the Alaskan coast. The trek was fraught with problems, including atrocious weather, predators, biting insects and passing caribou stealing reindeer. At times conditions forced a change in route. Two years eventually became five years, but finally 2,375 reindeer arrived at their destination; only 10 per cent of them were the original animals that set off on the trek.

These reindeer became the basis of the Canadian Reindeer Project, which for three decades was based at Reindeer Station on the east branch of the Mackenzie River and supplied reindeer meat for the region. The government project was eventually discontinued and the herd was privatised in the 1960s. Last summer I met the man who now manages the herd, Lloyd Binder, and he confirmed that the reindeer were still there and migrated every year to Richard's Island for summer grazing. The herd today is still used for meat production and when they cross the Mackenzie River in spring the spectacle has become a popular event for visitors to the region. Reindeer are actually very adept at swimming and it is common for herds to cross fast-flowing rivers, large lakes and the sea when they spend the summer on islands. Their large flat feet are very effective paddles and their hollow hair gives them tremendous buoyancy.

In 1990, we moved part of our herd of reindeer to a new site, on the Glenlivet Estate beside Tomintoul, some 25 miles as the crow flies from their Cairngorm Mountain home. This was a wonderful opportunity for us to 'spread our wings' and take some grazing pressure off the original site. A few male reindeer were the pioneers and they seemed to settle in quickly, quite at ease with their new surroundings. As it became obvious that the second site was a success, we introduced some females to the new area, one of which was our wily old reindeer Lilac. Within days Lilac had gone missing and we assumed she must be investigating the further reaches of the Cromdale Hills where we had put her and the other females. It didn't even cross our minds that she may be looking further afield! Days and then weeks went by and still there was no sign of her. And then much to everyone's surprise, she turned up on the Cairngorms at the normal 11 a.m. feed time. She must have crossed roads, fences, rivers and unchartered territory to get home. We never tried to put her back to Glenlivet – she had made her point!

However, the most incredible journey and one of the most bizarre stories of reindeer on the move must be the account of a young reindeer living in a submarine during the Second World War. The tale begins in the Russian port of Polyarnoe, situated just downstream from Murmansk on the Kola peninsula. HMS *Trident* docked at Polyarnoe during the war with engine troubles, and the commanding officer, Commander Geoffrey Sladen DSO, DSC, was invited to dine with the Port Admiral. Over the course of dinner, inspired by the amount of snow lying about, Commander Sladen spoke about the problems his young wife faced pushing their baby's pram up the steep hill to their house in snowy conditions. The Port Admiral suggested that what she needed was a reindeer.

A few days went by and nothing more was said, but just before the submarine shipped out to sea once more a reindeer arrived on the dockside, accompanied by a barrel of freshly picked lichen. Unwilling to 'look a gift reindeer in the mouth', Commander Sladen agreed to the animal coming on board. This was by no means an easy feat, as the reindeer wriggled and squirmed as they manoeuvred her down through the narrow torpedo-loading hatch, which had a diameter of just 21in.

Petty Officer James Riddoch was instructed to look after the 'gift'. He commented later that he had been on many courses during his time with the Navy but never had the topic of livestock care been covered. He had to judge for himself how much lichen the young reindeer should get each day and what other food they might be able to feed her. Naturally he soon won the nickname 'zoo attendant'.

On leaving Polyarnoe, HMS *Trident* rounded the North Cape of Norway to be met by hurricane-force gales. It was a particularly rough passage, and the reindeer sought refuge in the officers' toilets. However, the pitching and rolling of the submarine did not affect her appetite and she was soon causing chaos by trotting through the narrow gangways in search of the barrel of lichen.

A week into their voyage home, HMS *Trident* received orders from the Admiralty to stay on patrol as German warships were in the vicinity. As the days passed, the young reindeer began to acclimatise to her new home, always finding her way to the control room at the time when the main hatch was opened and fresh air poured in. She would stand with her front feet high on the ladder, taking in deep breaths of fresh air. When the conning tower was shut, the increasing staleness of the air often caused the reindeer to pant. A bigger problem, though, was that her lichen supply was dangerously low and would very soon run out. There was much discussion among the crew as to what they might feed her. However, she solved the problem for them by making her way past the officers' kitchen to help herself to leftovers from the waste buckets. In particular, she developed a taste for diluted condensed milk. Indeed, the little reindeer settled in well to life in the submarine, taking up residence in the captain's cabin when the submarine was submerged and heading for the main hatch for fresh air whenever it surfaced. She seemed to enjoy her life under the ocean waves, but the crew wasn't quite so sure. She had developed a pungent odour that, added to the normal smells in the submarine, made life almost unbearable when they were submerged for long periods.

A whole month passed, during which time the reindeer encountered plenty of wartime action, including the successful attack on the German heavy cruiser *Prinz Eugen*. When the submarine eventually arrived at Blyth in Northumberland, the reindeer (now known as 'The Goat' by the crew) had put on weight and her antlers had grown so much that they had to saw off the tips to get her back out through the hatch. To assist her exit from the submarine, they inserted her into a large canvas bag. Once on land she sprang out of the bag and legged it along the jetty with the *Trident*'s crew and dockworkers in hot pursuit. She was finally caught and led off to naval headquarters. She was named Pollyanna by the naval authorities at Blyth and eventually was presented to London Zoo, where she became quite a character. It was rumoured that whenever she heard bells or anything that sounded like the submarine's tannoy, she would lower her head as though she was preparing for diving stations!

· 14 ·

SWEDISH LAPLAND

AS I SAT on the east side of Lake Akkajaure in Swedish Lapland and gazed across the water to Sarek National Park and the summer grazing for many Swedish mountain reindeer, I could understand why Mikel Utsi had taken reindeer from his homeland in Arctic Sweden to the Cairngorm Mountains in the Highlands of Scotland. The two landscapes are remarkably similar – mountains with gentle slopes and high-altitude plateaus. We were there to meet up with Jussa Utsi, one of the late Mikel Utsi's nephews, who was going to take us by boat to his summer dwelling across the lake. We weren't quite sure what to expect but we had come laden with Scottish hospitality in the form of malt whisky, as this always seemed to go down well.

When Jussa arrived he seemed to be in no hurry to get over the water, despite the fact that it was quite late in the day. But then, this was the season of the midnight sun and the sun would remain high in the sky all day and all night. We had timed our visit perfectly. Jussa was going to his summer camp, Vaisaluokta, that day and would stay there for the next ten weeks. Close and extended family would join him during that time to help with reindeer calf marking, fishing, moose hunting and finally the reindeer bull slaughter in the autumn before the rutting season, when the bulls would lose their condition in pursuit of the cows.

While we sat waiting for Jussa to gather all his belongings, a steady stream of helicopters were landing and taking off. Their passengers were reindeer herders, who, like Jussa, were going out to more remote settlements in the mountains. Being a helicopter pilot looked like a pretty lucrative job in North Sweden. Today's mountain reindeer herders relied heavily on helicopters, snowmobiles and scrambler motorbikes – a far cry from reindeer sledges and the pack reindeer of yesteryear.

The lake we crossed in Jussa's boat was a flooded river valley, part of a massive hydroelectric scheme that destroyed vast areas of valuable reindeer grazing when the valley was originally dammed. Jussa, along with many Sami, was very bitter about the impact the hydroelectric scheme had had on their lives. They received no compensation for the loss of grazing and benefitted little from the electricity generated, which was primarily to meet the demand from urban areas further south.

As we crossed the lake and approached the far shore, we could see all along the shoreline small isolated family encampments nestled in the dense birch woodland. Each family had its own little cluster of buildings, with winding paths and raised wooden walkways across boggy areas linking the communities. Depending on the water level in the lake, which was governed by the engineers operating the dams, the unloading of the boat was a greater or lesser chore. The ugly eroded shoreline all the way round indicated that the level was low, so we had to carry the luggage that much further. Jussa begrudged any carrying between boat and dwelling, and recalled that on three occasions his family had had to move their houses on to higher ground to prevent them flooding as the water levels rose.

Before we left for Vaisaluokta we had been warned by a number of people about the ferocity of the mosquitoes, the 'Sami army' as Jussa described them. Luckily the hot dry day and high-speed boat ride had kept them away. However, as the day cooled, the mosquitoes became more evident and we were very glad of the mosquito screens across the windows. When we asked Jussa about the mosquitoes, he extolled their virtues. 'Mosquitoes,' he said, 'are the reindeer herders' best friend. Biting mosquitoes help to keep the reindeer together, which helps us to both locate the herds and gather them into the corrals for calf marking.'

When most of us think of black pudding as a sausage made from blood, the Sami go one step further in the culinary delights of reindeer blood and make blood pancakes, substituting milk with fresh blood. They are very tasty — I've tried them!

He added that they also keep the tourist population away, leaving the mountains undisturbed for the reindeer to graze in the summer.

That night we dined on a native fish, Arctic char, freshly caught in the lake and cooked over a smoky fire, dried reindeer meat and freshly baked bread. The reindeer meat came in the form of a whole shoulder, still on the bone, which had been salted and dried. All Sami carry a sharp knife, which they use to cut off slices of meat.

The life of Swedish Sami had changed dramatically over Jussa's lifetime. He remembered as a child travelling with herds of reindeer in spring on sledges pulled by reindeer as they moved many miles from their winter grazing in the forest to summer in the mountains. The journey took many weeks, with some cows calving on the way. The whole family made the journey, the fit and able on skis, the very young and the old traveling in sledges pulled by reindeer. Today the reindeer still migrate but the herds are guided by a few Sami on snowmobiles. The journey time is much shorter and fewer people are involved. When asked if he yearned for the traditional ways, Jussa answered emphatically. 'Life was very hard then,' he said. 'Modern-day reindeer herding is better for everyone.'

The next morning we headed up on to the ridge behind the Utsi encampment. As we climbed up through the trees along a well-worn path we were impressed by the wealth of vegetation. After a winter shrouded in snow and ice, the lengthening days gave the plants ample opportunity to grow at an amazing rate. As we gained height we noticed that the tree layer continued to grow profusely, but in a more shrub-like form. In fact, the shrubbery was so dense that it was almost impossible to walk off the path. We could understand how the reindeer here were able to grow so quickly over the short growing season, laying down sufficient fat reserves to last them through the lean winter months.

Our walk took us on to the ridge, where there were several permanent corrals with long guide fences radiating away from them. This was where the reindeer were gathered for calf marking in the summer and the bull slaughter in the autumn, before being moved down off the mountains for the autumn rut and winter in the forests. Remnants of previous years' activities abounded: bones, skulls and antlers and even the odd beer tin littered the area. The corrals were pretty makeshift, formed out of thin poles, bleached by years of winter storms, and disheveled netting wire. Each year, cursory patching took place to keep the fences intact, but no new fencing had been erected for a long time. Although there were no reindeer close by, with binoculars we could see groups of a few hundred gathered together on the snowfields. In the heat of the day this was a good way for the reindeer to cool off and avoid being bitten by mosquitoes.

When we returned, Jussa was at the side of the lake talking on his mobile phone. He had a helicopter delivery organised and he was down there waiting for its imminent arrival. As the helicopter flew towards us over the lake, we could see something dangling below it. The cargo turned out to be two scrambler bikes that would be used in the mountains to gather the reindeer for calf marking in a few days' time. Unfortunately our time was short, so we could not stay for that. Luckily, though, on our way up to Vaisaluokta we had been able to visit a calf marking with Forest Sami further south.

Our guide and contact was a Swedish professor of reindeer husbandry, who was particularly interested in comparing the growth rates of different populations of reindeer in relation to climate and vegetation. We arrived in Arvidsjaur at about midday and went straight to the house of one of the Sami elders, who was organising the catch that day. We were told to go and find somewhere to stay for the night and they would call us once they had got the herd in. I suspect we would have been more of a hindrance than a help when it came to the tricky task of getting reindeer into a corral from the surrounding forest. The day was wearing

on, so, having found suitable accommodation, we headed into town for supper. Back at our sleeping quarters for the night, we reckoned that since it was getting late, they must have been unsuccessful in rounding up the reindeer and the event would be postponed to another day. But we were wrong! A short time later the phone rang and we were told it was 'action stations'. The reindeer had been corralled and we should go immediately to the site.

The sight that met our eyes was almost unbelievable and the noise was deafening. In the midst of a continuous coniferous forest was a corral the size of a football pitch, made up of propped-up wire netting and hessian. Inside this makeshift fence were at least 3,000 reindeer, all trotting round, the cows and calves constantly grunting to one another. It was midsummer so all the adults were growing their new velvet antlers, and the bulls' antlers were distinctly longer than those of the cows. They were all still in full winter coat. In the midst of this moving mass of reindeer, numerous people stood around looking intently at the reindeer. Each calf had to be individually identified by correctly reading its mother's ear notches. This was no mean feat when you consider that any individual may have as many as four or five different notches in each ear to determine ownership. In addition, reindeer ears are not very big and are extremely hairy, and the reindeer were constantly on the move. Once a calf was matched to a mother it would be caught by the back leg, using a sliding loop on the end of a long pole. Then a knife was used to cut ear notches identical to the mother's before the calf was released back into the group. A piece of ear from each calf was kept to count the calves marked at the end of the night's work. The constant grunting was mostly generated by mothers and calves trying to relocate each other in the swirling mass of beasts. The reindeer always moved in one direction, anti-clockwise. The herders told us that the reindeer would only go clockwise when they were panicking, which might result in a stampede out of the makeshift corral. As long as the reindeer moved anti-clockwise they were calm and there was no problem.

Inevitably, this process was very slow and with over a thousand calves to mark it was going to take hours. By about 4 a.m. the majority of the calves had been marked. Towards the end the people with the long sticks systematically caught the last few unmarked calves and hung a large number round each one's neck, then nearly everyone left the corrals and sat round open fires drinking tea and eating dried reindeer meat while the reindeer settled. A couple of men were left with the reindeer, writing down the calf numbers and trying to match them to their mothers. Then the catchers returned, caught the numbered calves and cut the corresponding notches in their ears. By 7 a.m. the last of the calves was marked, and the whole group was released back into the forest. The reindeer could finally relax and everyone could get some sleep. Later that morning we went back to the catch-up site. Trampled ground, a few small pieces of calves' ears and a couple of reindeer aimlessly wandering around was all that was left from the frantic activity of the night before.

In hindsight, it was of course the obvious time of day to handle the reindeer. Although the sun still shone, it was much lower in the sky at night and so much cooler. Reindeer at the end of June are still in full winter coat so any handling during the heat of the day would have undoubtedly caused them stress. Also the mosquitoes were much less active in the middle of the night. It was a relief to know they also had body clocks. Perhaps that's the answer for tourists visiting Lapland in summer: become nocturnal, as it seems the Sami are.

Some years later Alan and I were lucky enough to return to the mountains above Vaisaluokta in the autumn where we stayed with another of Mikel Utsi's nephews, John Erling Utsi, at his summer cabin at Kutsjaure, beside the lake. We took the 'helicopter taxi' into this remote corner of the Sarek National Park and spent the next few days exploring the mountains, getting involved in a moose hunt and watching the gathering of a huge herd of reindeer for the bull slaughter at the same mountain corrals we had wandered round in the summer years before.

Autumn colours in Lapland are spectacular with the leaves of the birch trees, larch trees and low-lying shrubs changing from green to every shade of red, yellow and brown before they fall to the ground for the winter. And there are frequent dustings of snow reminding us that winter isn't far round the corner. After a summer of continuous light, and for the reindeer unlimited grazing, they are looking at their very best and the bulls have grown their large impressive antlers ready for the rut. The reindeer are also in their autumn colours, with the new lighter-coloured winter coat already growing through darker summer coat. It is this time of year that the first slaughter of adult reindeer takes place: it is also the time of year for moose hunting, another important source of meat and income for the Sami.

Moose are the biggest of the deer species, a mature bull moose standing 2m or more at the shoulder. They are solitary, live in dense forest and are, despite their size, incredibly difficult to see, even at close range. With dark-brown bodies and light-coloured legs they blend perfectly into the birch woodland. It can take many hours to both locate a moose to shoot and then extract the carcass, which can weigh anything up to 400kg. On the day we went out with John we spent many hours spying a hillside of birch trees across the lake. Once we had located one, John headed across the lake in his small boat, landed on the shore and crept up through the wood. A single shot confirmed he had been successful and we headed across to help John extract the carcass. I never imagined quite how big moose were: this young bull moose was the size of a small horse. So just how would we manage to get the carcass back to base? It was getting late and after bleeding and gutting the moose John placed a fine mesh net over the body, to prevent other predators taking their share overnight. Already, as we were leaving, birds of prey were circling overhead aware that there was something tasty to eat below. Indeed, the following morning when we went back to retrieve the carcass I had the best sightings of both golden and white-tailed eagles that I had ever had in my life. Luckily there was a bit of a slope down to the shore and in the morning we managed to drag the moose down to the boat; John then cut the carcass up into manageable pieces and we placed them in the boat. We certainly needed the helicopter to take our extra cargo back at the end of our stay!

John's summer cabin, although off the beaten track, is not far from the long-distance walking route that stretches the length of Sweden, 'the

Kungsladen', and Alan and I walked part of this route on one of the days we went exploring. Over the boggy ground were wooden walkways and as we crossed them we noticed under the walkways were lots of lemmings.

Although reindeer are herbivorous they have been known to eat lemmings, small arctic rodents, during the wintertime. With depleted reserves and food, reindeer are certainly 'on the scrounge' for anything edible at this time of year.

Lemmings are a very prolific small mammal found in the Arctic and being low down in the food chain makes them a popular prey species. The manmade walkways have provided the lemmings with a hideout from circling birds of prey and ground predators like foxes. During winter, snow provides a new hiding place and they store grass in the summer, eating the dried vegetation under the snow during the winter months.

Quite by chance we then stumbled upon a traditional reindeer calving ground. We knew this because pregnant females usually cast their antlers when they calve and the ground was littered with cow antlers. It was a reindeer enthusiast's paradise; some of the antlers we picked up must have been lying for years because they had lichens growing on them. Female reindeer antlers are much smaller and more delicate than the mature bull antlers but on first glance could be easily confused with young bull antlers, which are of similar dimension. However, there is a fundamental difference between the two and this is at the point of casting at the base of the antler. Female antlers are concave at the base and males convex. So even in archaeological digs the sex of reindeers found can be determined by whether the antlers are convex or concave.

The deserted corrals that we had visited in summer time many years before were to be transformed into a hive of activity. From as far as the Norwegian border Sami reindeer herders had been out for days scouring the mountains, in helicopters, scrambler motorbikes and on foot to gather the reindeer. This was a community effort like gathering hill sheep, to bring the reindeer in to their mountain corral where the herders would find out how many calves had survived the summer, what sort of condition the reindeer were in and to mark any calves that slipped the net earlier in the year. It was also an opportunity to take the first slaughter of any adults: castrates and old females without calves.

It was amazing to watch thousands of reindeer on the move across the mountainside, groups coming in from all different directions, presumably along routes they had traditionally taken for hundreds, maybe thousands of years. As they were chivvied along by herders on motorbikes and in helicopters, we watched from the corrals. The reindeer were like a swarm of insects – hard to see without binoculars. And then as they came closer we could hear the clicking of their tendons, the grunting between mothers and calves and finally the reindeer surged into the corrals, fanning out after passing through the gateway and finally settling in the large enclosed area. The work was done for that day; tomorrow the men and women would return to sort them out before sending them on their annual journey off the mountain before winter sets in and into the forests for the rut and finally the winter. For the next few weeks the reindeer would be on the move, finding fresh grazing and delicious mushrooms while making their way down the Lulea valley to their wintering grounds in the forests around Porjus, Jokkmokk and Murjek.

• 15 •

THE FROZEN NORTH

OUR DESTINATION WAS the Yamal-Nenets Autonomous region; the flight took three hours from Moscow due east and crossed two time zones and the Ural Mountain range. As the aeroplane descended towards the town of Salekhard all we could see was snow, frozen rivers and swamps, and large open spaces, interspersed with a thin spattering of spruce trees. Although the area is rich with deciduous woodland too, from the aeroplane these trees were not visible to the eye as they had no leaves on them at this time of the year. This is the northern edge of the taiga, eventually stretching north onto the Arctic tundra.

Salekhard is the only town in the world that is actually situated right on the Arctic Circle and in true Russian style there is a huge shiny monument celebrating its claim to fame. It is built on a tributary of the great River Ob, which begins its journey in South Siberia snaking some 3,600km north finally reaching the Kara Sea at the Yamal peninsula. Salekhard originally made its riches from the fur trade. It had a gruesome history of labour camps during Soviet times and today it's fortunes are in oil and gas – 95 per cent of Russia's reserves can be found in this area.

For the people of Salekhard it's always a long winter, temperatures plummet in October, the rivers are frozen until the following June and snow is everywhere apart from the runway at the airport and the roads in the city. The 50,000 inhabitants of Salekhard never venture far from the town, or even their houses, in winter, and in summer the area is awash with mosquitoes. The indigenous people, the Yamal Nenets, don't venture into town except for shopping, socialising and health care. The rest of the time they are on the move with their reindeer, living all year round in chums (large conical tents), guided by their reindeer's need for grazing and to avoid biting insects.

Our trip to Salekhard had been organised by the World Reindeer Herders Association and coincided with a 'reindeer people's festival'. Festivals like this occur all over the Russian Far North during the winter. It lasts one or two days, reindeer herders gather to sell their wares, show off their reindeer and take part in various competitions, including reindeer racing, lassoing, wrestling and best-dressed child. In Salekhard it is the only occasion when the townsfolk get an opportunity to meet the nomads. They can pay for a ride on one of their sledges pulled by a team of reindeer, and sample raw meat and fish in one of their chums.

For a reindeer enthusiast like myself I was in awe of the whole spectacle. At least a dozen chums had been erected, each one 6m high with a floor diameter of 9m. Each tent needed at least eighty reindeer skins sewn together to complete them and inside they were cosy warm with a wood-burning stove chugging away. Close by the erected tents were dozens of teams of reindeer, harnessed in threes, fours and fives to functional-looking wooden sledges. Some of the sledges and reindeer had simple harness and trappings; others had ornate cloth draped over the sledge and intricate harness and halters for the reindeer. Despite loud music banging away up on the main stage, people milling around and snowmobiles zooming to and fro, the reindeer were completely placid, heads down waiting patiently for their 'driver' to goad them into action. A team of reindeer were loosely held together with leather straps, the lead reindeer, always on the left, slightly ahead of the rest. To set off, the driver would lead the front reindeer forward, pulling the back ones with him and then when the moving sledge came broadside to the driver, he would jump on, further encouraging the reindeer to go straight ahead with a 'nudging' pole about 2m long which reached the head of the front reindeer. Reindeer normally circle anti-clockwise and so the long pole is always on the left to guide the team of reindeer straight ahead. To stop the sledge, the driver pulls the lead reindeer with the single left-hand rein and brings the team to a halt as they circle round. Despite plenty of teams coming and going I never witnessed two teams getting tangled up: they were amazingly adept at avoiding each other.

As well as the more sedate reindeer and sledge rides, there was the highly competitive reindeer racing: the same principle but with more action. Each race involved two teams and with the competitors raring to go on the start line they set off amidst a clatter of hooves and whooping from the driver, who with great dexterity jumped on to the passing sledge as they set off at breakneck speed. What happened next varied hugely. Some teams took a linear path heading for the far end before returning, probably faster than they left. Reindeer are very social animals and for them the quicker they returned the better. However, many a team was less direct, the lead reindeer veering sharply to the left, despite the driver trying to persuade him otherwise and the whole lot ending up in a heap on the sidelines in the deep snow. No doubt the racing reindeer were hand picked for the job and were the strongest and fastest of their draught reindeer. Normally, castrated males are used for this task. Here I was surprised to see some of the teams also included females and young bulls. Nearly all of them had their antlers cut off. I suspect the bony antlers would get in the way when harnessing and in any case there was probably a market for the antlers too.

The traditional dress of the Nenet is reindeer skin, elaborately embroidered with colourful felt edging and shiny beads. The women and children are the most ornate, their outer garment covering them from head to toe, with sewn-in reindeer skin gloves. Clad in these incredibly warm coats, the small children were conical shaped, arms stuck out at an angle and rocking from side to side as they walked along. The men wore long reindeer-skin chaps or leggings but their tunics were made of thick cloth. I suppose the difference in dress was due to how they travelled and went about their daily chores: the women and children ride on the sledges while the men would often walk or ski alongside the caravans of reindeer and the men would be more actively involved in lassoing the reindeer for harnessing up.

For me, the most incredible part of the whole process of living in the Yamal, which for the Nenets means 'the edge of the world', was how they managed to dismantle their tents, pack them with all their belongings onto the sledges and then move to their next stop. This process requires many teams of trained reindeer to transport the equipment and is repeated time and time again throughout the year, as they travel north in the spring and south for the winter up and down the Yamal peninsula. Not to mention the many hundreds of 'untrained reindeer' that are herded along with the nomadic families to the next, transient destination. Teamwork at its best!

During the last thirty years these people, who herd approximately 760,000 reindeer, have been massively affected by the intrusions of fossil fuel exploration. However, they have also adapted and in many cases benefitted from the infrastructure that has been developed alongside extraction and transportation of oil and gas. The melting of the Arctic sea ice (because of global warming) is being seen by the Nenet as a future business opportunity, as the opening of the northern seaways means they can reach out to new markets for their reindeer meat.

While in Salekhard we visited the local museum of natural history and indigenous people. On entering the museum we were greeted by a complete skeleton of a woolly mammoth, 3.5m high at the shoulder and resplendent with immense ivory tusks. These extraordinary animals died out during the retreat of the last Ice Age, some 11,500 years ago, and to this day their remains are still being discovered in the permafrost. Beside this giant was a glass cabinet displaying a baby mammoth named Lyuba after the wife of the Nenet reindeer herder who found her in 2007 on the bank of the River Yuribei. Completely intact and just 85cm tall, 130cm long and weighing 30kg, Lyuba is thought to have been just 35 days old when she perished 42,000 years ago. When discovered, this incredibly well-preserved body caused ripples in the world of palaeontology, travelling to various countries to be examined, displayed and finally

returned to the place she lived all those thousands of years ago.

Large mammals like woolly mammoth, hairy rhino, cave bear and cave hyena all died out at the end of the last Ice Age, but the reindeer and caribou survived the final retreat of the glaciers to their present distribution in the Arctic and sub-Arctic. By 10,000 years ago North America had lost more than 70 per cent of its mega-fauna, including the mastodon and sabre-toothed cats. In Europe a similar loss was experienced, but not on such a large scale; 30,000 years ago large mammals like mammoth, woolly rhino and the giant deer megaloceros were abundant, but by 10,500 years ago they were entirely extinct.

These immense extinctions took place over a long period of time and there are a number of theories as to how they came about. An inability to adapt to changing climates and overhunting by prehistoric man are both thought to have contributed, and it was the larger, slower-breeding species that succumbed. The cold fauna of today is represented by animals that are confined either to Arctic regions, like reindeer, musk ox, lemming and Arctic fox, or to the steppe grasslands of Eastern Europe and central Asia, like the saiga antelope and ground squirrel. These faster-breeding animals are more adaptable to change. They reach sexual maturity at a young age and have a relatively short gestation period, so they were able to react faster to changing environmental conditions.

Back home in Scotland, the recent spring snowstorm had abated and the access road to Cairngorm was clear of snow, allowing us to get out onto the hill to find and feed our reindeer. Calling them from afar, the herd approach in a long line through the deep snow, following in each other's footsteps. As they walk, their back feet are placed in the prints of the front feet, so lessening the tendency for all four feet to sink into the snow, the front foot having already compacted the surface for the back foot. Once the herd meets us they patiently wait, silent but attentive, for the food to appear. There is no incessant noise like a hungry sheep baaing for its food. Making a noise is costly, because by expelling air to grunt the animal loses heat. Reindeer grunt as their means of communication but grunting is kept to a minimum during the winter months.

However, reindeer and caribou are extremely sociable animals and prefer to stay together. In low visibility, blizzard conditions or thick fog they can locate each other by their clicking tendons. As the reindeer articulates its foot, a tendon in the joint above slips across, producing a clicking noise. All reindeer and caribou click, and when a large number are on the move together the clicking is very noticeable. There is very little energy cost involved so the herd can remain in contact without losing valuable heat.

During the snow blizzard of the night before the reindeer would have faced the north wind, which keeps the individual hairs flat trapping air for extra insulation. With an unruffled coat, the blowing snow sticks to the forehead, leaving the reindeer with a facial icepack and frosted antlers. Although we like to find the reindeer daily in the winter and feed them, it is more a management exercise than a necessity for them to be fed. Reindeer show a decline in food consumption during the winter months. This reduced appetite, combined with a lower metabolic rate, means that long periods of the day can be spent laying up and resting, further increasing energy conservation. It is very common for reindeer to happily remain motionless for relatively long periods of time during the winter. Patience is second nature to them.

Feeding over, the reindeer wander away for a drink. They push their noses in the snow, scooping it up to eat! It would seem they positively prefer snow to water. This is very sensible, considering the abundance of snow and the lack of running water in the winter. The small amount of heat expended to melt snow for drinking water is worthwhile compared to the potential loss of body heat that would be incurred in searching for running water, and with a reduced intake of food and

The largest herds of wild reindeer in Russia are in the Taimyr Peninsula. However, their numbers have declined sharply from 1 million in 2000 to only 60,000 today. As one of the most researched herds in the world over the last fifty years, climate change and industrial development are thought to be the main causes of their decline.

water there is less toxic waste to excrete, so another small contribution to reducing heat loss is made by urinating less.

These adaptations are all small, but added together are hugely significant ways in which reindeer can survive in their Arctic environment. Which seems ironic when you think of this highly adapted animal living in such an extreme environment by conserving its energy, alongside the ever-increasing efforts by man to extract as much 'energy' from the Arctic, in the form of oil and gas, as possible. The Arctic is the 'canary in the coal mine' – the first area in the world to alert man to the effects of climate change and global warming.

Melting permafrost, unexplained 'sinkholes' in the tundra, vanishing pack ice, rapid freeze/thaw of snow and invasions of insects more associated with more southerly climes are all effects of human-induced climate change. One wonders what the future holds.

CAIRNGORM REINDEER HERD

The Cairngorm Reindeer Herd is Britain's only free-ranging herd of reindeer found in the Cairngorm Mountains in the Highlands of Scotland. These tame and friendly animals are a joy to all who come and see them. Reindeer are not just for Christmas! There are currently around 150 reindeer in the herd, mostly ranging on the Cairngorm Mountains with the remainder on the Glenlivet Estate, the locations being some 30 miles apart.

The Centre in Glenmore is open from the February half-term holidays right through the year to early January. Daily Hill Trips with one of our experienced reindeer herders guide visitors up onto the mountainside where the main herd live all year round. The reindeer's soft velvet noses mean they are a delight to hand-feed and, with their endearing and friendly nature, visiting them is an unforgettable experience. During the winter, the trip may depend on whether we can find the herd – we said they're free-ranging and we mean it!

For those unable to attend a Hill Trip, the paddocks and exhibition in Glenmore offer an alternative easier way to see the reindeer. These are open daily and are perfect for children and for visitors who are unable to go up the hill. They are wheelchair and pushchair accessible, and as well as being able to see the reindeer, there is lots of information about every aspect of our reindeer. Children will also enjoy exploring Santa's Bothy and the Elves' House, and finding the answers to our quiz. On December weekends leading up to Christmas we also run 'Christmas Fun' with extra crafting activities and visits from Santa.

All the reindeer in the herd have a name and are individually recognisable. Our adoption scheme, established in 1990, helps to maintain their enviable lifestyle on the Cairngorm Mountains. Adopters become valued supporters and receive a certificate and photograph of their chosen reindeer, additional information on reindeer and souvenirs, along with two newsletters over the year. They can also visit the herd for free! The adoption subscriptions are spent entirely on the upkeep of the reindeer including their food and welfare such as medicines and veterinary care.

Over November and December, our tame and tractable adult male reindeer go out and about on tour nationwide, increasing public awareness of the herd and raising funds to help to maintain their lifestyle. Each individual reindeer will only do a couple of weeks' work before returning back to their mountain home, where they can rest and relax for the rest of the year.

Photographs: Tilly and the Cairngorm herd. Courtesy of John Paul.

Websites

www.cairngormreindeer.co.uk
www.reindeerherding.org

REINDEER FAMILY TREE

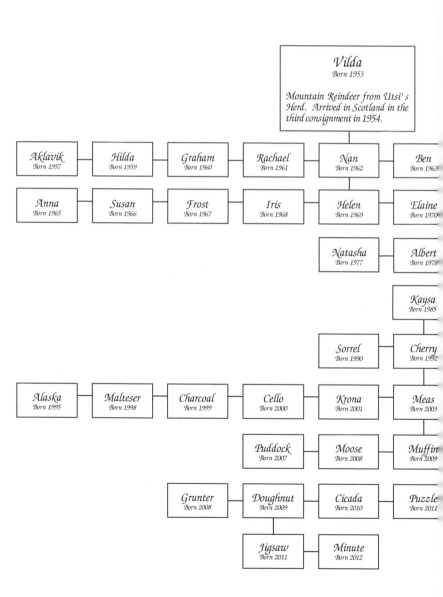

Vilda
Born 1953

Mountain Reindeer from Utsi's Herd. Arrived in Scotland in the third consignment in 1954.

Aklavik	Hilda	Graham	Rachael	Nan	Ben
Born 1957	Born 1959	Born 1960	Born 1961	Born 1962	Born 1963

Anna	Susan	Frost	Iris	Helen	Elaine
Born 1965	Born 1966	Born 1967	Born 1968	Born 1969	Born 1970

Natasha	Albert
Born 1977	Born 1978

Kaysa
Born 1985

Sorrel	Cherry
Born 1990	Born 1992

Alaska	Malteser	Charcoal	Cello	Krona	Meas
Born 1995	Born 1998	Born 1999	Born 2000	Born 2001	Born 2003

Puddock	Moose	Muffin
Born 2007	Born 2008	Born 2009

Grunter	Doughnut	Cicada	Puzzle
Born 2008	Born 2009	Born 2010	Born 2011

Jigsaw	Minute
Born 2011	Born 2012

Micky *Born 1964*	**Jemima** *Born 1965*	**Sarah** *Born 1966*	**Wind** *Born 1967*	**Wendy** *Born 1968*
Holly *Born 1971*	**Hawk** *Born 1972*	**Nadia** *Born 1973*	**Feshie** *Born 1974*	
Arran *Born 1979*	**Morag** *Born 1980*	**Erik** *Born 1985*	**Torrent** *Born 1986*	
Till *Born 1989*				

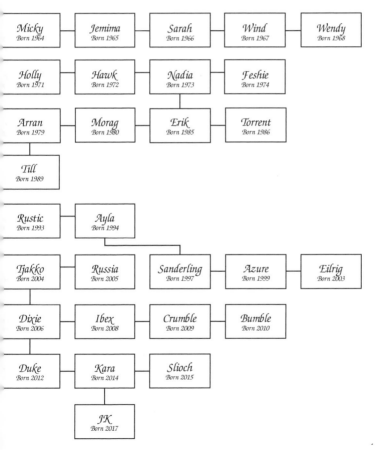

Rustic *Born 1993*	**Ayla** *Born 1994*			
Tjakko *Born 2004*	**Russia** *Born 2005*	**Sanderling** *Born 1997*	**Azure** *Born 1999*	**Eilrig** *Born 2003*
Dixie *Born 2006*	**Ibex** *Born 2008*	**Crumble** *Born 2009*	**Bumble** *Born 2010*	
Duke *Born 2012*	**Kara** *Born 2014*	**Slioch** *Born 2015*		
	JK *Born 2017*			

PLAYING
WITH
FIRE

THE ART OF CHOPPING
AND BURNING WOOD

PAUL HEINEY

978 0 7509 7994 8

The History Press

The destination for history
www.thehistorypress.co.uk